Severe Plastic Deformation

This book discusses in detail grain refinement through various severe plastic deformation processes such as equal channel angular pressing, high-pressure torsion and twist extrusion. It also deals with the grain refinement mechanism, the influence of process parameters on properties of fabricated micro/nano material and thermal stability as well as commercialised nanostructured metals and alloys produced by severe plastic deformation techniques and associated challenges.

Key features:

- Includes individual severe plastic deformation processes and their engineering features in detail.
- Discusses commercialised nanostructured metals and alloys produced by severe plastic deformation and challenges.
- Deals with the grain refinement mechanism and the influence of process parameters on the properties of fabricated micro/nano material.
- Explores pertinent functional properties.
- Gives a comprehensive view of the successful application of severe plastic deformation.

This book is aimed at graduate students and researchers in materials science, mechanics and manufacturing engineering.

Severe Plastic Deformation

Practices and Properties

B. Ravisankar and P.C. Angelo

CRC Press
Taylor & Francis Group
Boca Raton London New York

CRC Press is an imprint of the
Taylor & Francis Group, an **informa** business

Designed cover image: B. Ravisankar

First edition published 2026
by CRC Press
2385 NW Executive Center Drive, Suite 320, Boca Raton FL 33431

and by CRC Press
4 Park Square, Milton Park, Abingdon, Oxon, OX14 4RN

CRC Press is an imprint of Taylor & Francis Group, LLC

ISBN: 978-1-032-97705-8 (hbk)
ISBN: 978-1-032-97917-5 (pbk)
ISBN: 978-1-003-59611-0 (ebk)

DOI: 10.1201/9781003596110

Typeset in Times
by codeMantra

Contents

Preface

Metallic materials undergo plastic deformation on the application of sufficient stress. When we try to deform them severely, it leads to their failure. However, it is possible to extend their ability to plastically deform by applying deformation restoration mechanisms such as dynamic recovery or dynamic recrystallisation, which help them to modify their microstructure. Metallic materials with FCC and BCC crystal structures can be deformed to a large extent at ambient temperatures, since they have sufficient slip systems, while those having HCP crystal structures, having limited slip systems, need to be deformed at high temperatures so that additional slip systems can become active. As per the Hall-Petch relation, the strength of a material varies with grain size. To achieve high strength in an alloy system, it is necessary to refine its grain size by subjecting it to a thermomechanical treatment so that fine grains can result due to dynamic recrystallisation. However, grain sizes thus obtained are only in the range of 1–10 microns – not sufficient to increase the strength sufficiently. A recent method of refining the grains to submicron ranges is to subject the material with an initial coarse-grained (CG) (>100 mm) microstructure to severe plastic deformation (SPD), in which, by a series of plastic deformation steps, it can be refined to an ultrafine-grained (UFG) microstructure. SPD is achieved when a material is plastically deformed repeatedly without a net change of shape. While the process is repeated, the plastic strain in effect gets added with each cycle, resulting in a large achieved strain. Unlike the conventional plastic deformation techniques, such as forging, rolling and extrusion, what distinguishes SPD is the application of high plastic strain magnitude with high rates so as to attain SPD. Subjecting a material to SPD refines its microstructure, increases its yield strength (YS) anywhere from three to five times and reduces its work-hardening ability. Materials that have undergone SPD exhibit interesting and unique properties not usually observed in conventional CG and mildly deformed materials.

Bridgman defined the process of SPD in the 1950s as a method of metal forming under an extensive hydrostatic pressure that is used to impose a very high strain on a bulk solid without introducing any significant change in the overall dimensions of the sample but having the ability to produce exceptional grain refinement. However, only in the last three decades, metals and alloys with ultrafine grain and nanostructures have been studied due to their unique properties.

Among the various methods to produce materials exhibiting ultrafine grain (UFG) and nanostructures (NS), SPD has attracted a lot of attention as it can produce UFG and NS materials without porosity. After intensive investigation by a large group of researchers, SPD has been accepted as an important tool for improving the mechanical properties of metallic materials. The improvement of the properties by SPD methods is gained by significant refinement of the grain structure down to UFG or nanocrystalline (NC) regimes. UFG and NC materials display unique effects such as high strength combined with very good ductility, hardening by annealing, inverse temperature dependence of toughness and improved corrosion resistance not observed in their CG counterparts. During the last three decades, due to the interest in UFG materials, a number of SPD processes have been developed for various workpiece shapes such as sheets, discs, rings, pipes and bulk ones.

Our book on SPD discusses in detail the following processes among those developed:

- Equal channel angular pressing (ECAP)
- High-pressure torsion (HPT)
- Twist extrusion (TE)
- Cyclic extrusion and compression (CEC)
- Multi axial forging (MAF)
- Accumulative roll bonding (ARB)
- Groove pressing (GP)
- Friction stir processing (FSP)

We have also discussed grain refinement mechanism, influence of process parameters on the properties of the fabricated micro/nano material, thermal stability, commercialised nanostructured metals and alloys produced by SPD techniques and the challenges for commercialisation of nanostructured metals and alloys.

It is our fond hope that this book will serve as a textbook for courses in engineering and as a useful reference book for practicing engineers and that it will stimulate research and development in the important field of Severe Plastic Deformation in the coming years.

About the Authors

B. Ravisankar, Department of Metallurgical and Materials Engineering, National Institute of Technology, Tiruchirappalli, has more than 25 years of teaching and research experience. He has contributed a chapter on equal channel angular pressing (ECAP) to the *Handbook of Mechanical Nanostructuring*. His books *Powder Metallurgy: Science, Technology and Applications*; *Non-Ferrous Alloys: Structures, Properties and Engineering Applications*; *Periodic Table of Elements*; and *Introduction to Steel – Processing, Properties and Applications* have been well received as textbooks by numerous institutions.

P.C. Angelo, PhD, is internationally renowned in the fields of Powder Metallurgy and Materials Characterisation for over 40 years. He has taught, researched and published extensively in the areas of optical emission spectroscopy, optical microscopy, X-ray diffraction, X-ray fluorescence, electron probe microanalysis, scanning electron microscopy and scanning Auger microprobe. He began his career as a scientist at the Defence Metallurgical Research Laboratory (DMRL), Hyderabad, India. His books *Powder Metallurgy: Science, Technology and Applications*; *Materials Characterisation*; *Non-Ferrous Alloys: Structures, Properties and Engineering Applications*; *Periodic Table of Elements*; and *Introduction to Steel – Processing, Properties and Applications* have been well received as textbooks by numerous institutions. Presently, he is Visiting Professor at PSG College of Technology, Coimbatore, Tamil Nadu, India.

BOOKS PUBLISHED BY THE AUTHORS

1. P.C. Angelo and B. Ravisankar, *Non-Ferrous Alloys: Structures, Properties and Engineering Applications*, Cengage Publishers, 2018, ISBN: 9789387994041
2. P.C. Angelo and B. Ravisankar book on *Periodic Table of Elements*, Mahi Publishers, Ahmedabad, 2019, ISBN: 978-81-940137-1-6
3. P.C. Angelo and B. Ravisankar book on *Introduction to Steel – Processing, Properties and Applications*, CRC Press, Taylor & Francis Group, Florida, USA, 2019, ISBN: 9781138389991
4. P.C. Angelo, R. Subramanian and B. Ravisankar book on *Powder Metallurgy; Science, Technology and Applications*, PHI Learning Private Limited, 2022, ISBN: 978-93-91818-48-7

1 Introduction

1.1 INTRODUCTION

The developments in all important fields of micro and nanotechnology depend on the availability of suitable materials and successful shaping methods of engineering micro and nanocomponents required for building MEMS and nanodevices. The most challenging problem is in making many of the raw components for functional nanomaterials. Another major challenge is the lack of chemical, morphological, or mechanical stability in many micro- and nanomaterials. These often lead to spatial distortions, suboptimal thermal behaviour, reduced mechanical response and poor electrical properties, all of which undermine the overall system performance. An alternative approach is to employ bulk micro and nano-grained materials while constructing systems and devices at the micro and nano-scales. Besides employing for the fabrication of micro/nano components, bulk micro and nano-grained materials have enhanced mechanical and other properties. Hence, the bulk micro and nano-grained have replaced their coarse-grained materials in many applications. However, conventional metallurgy cannot supply materials featuring grains substantially smaller than the characteristic dimensions of the engineering micro-components.

Besides MEMS devices, modern engineering applications frequently require materials with a set of multifunctional, often conflicting properties. For example, enhanced mechanical properties need to be combined with improved physical (electrical, magnetic, etc.) and/or chemical (corrosion resistance, biocompatibility) properties, which have to be engineered and optimised simultaneously. These requirements have created a paradigm shift in which the classical materials selection approach has been replaced by the design of material microstructures to achieve certain performance requirements, subject to constraints on individual properties such as strength, conductivity, and corrosion resistance. Thus, production of bulk micro/nanostructured materials has become necessary. The need for bulk nanostructured materials finds applications ranging from hydrogen storage to construction engineering.

The processing and mechanical behaviour of bulk nanostructured materials have become the most interesting new fields of research on

DOI: 10.1201/9781003596110-1

advanced metallic systems. Many nanocrystalline materials possess near-theoretical strength and exhibit high values for fracture toughness, resilience, and fatigue resistance. There continues to be interest in these nanomaterials for use in biomedical, structural, and mechanical applications, and there are now a high number of research programmes worldwide.

1.2 INTRODUCTION TO SEVERE PLASTIC DEFORMATION

There are two main approaches to obtain bulk micro and nano-grained materials, called "bottom-up and top-down" approaches. In the first method, ultrafine powders consisting of nano-sized particles are synthesised via various techniques and then compacted into bulk form, using methods similar to those applied in conventional powder metallurgy (PM). Since the bottom-up production of bulk micro/nano-grained materials usually involves the synthesis of powder precursors as the first step, followed by consolidation at high temperatures, it is commonly called a "two-step" process. The bottom-up processed micro/nano-grained materials have much higher strengths than their coarse-grained (CG) counterparts, but at the expense of ductility. The poor ductility is suspected to be a defect-induced phenomenon. Probable causes are linked to volume defects, such as residual porosity, poor inter-particle bonding, and impurity contamination, introduced during either step of the "two-step" process. To avoid such complications, in the second top-down methodology, one starts with a bulk material composed of coarse grains and refines the grain size into the micro/nanograin regimes. Since no powder synthesis is involved, it is also called a "one-step" process, even though accumulative "passes" or "steps" are often needed to achieve the desired grain size levels.

The bulk coarse grained material is deformed severely so as to impose true strain of more than one. The amount of plastic strain produced by classical metal forming operations is often limited because of the material or tool failure. In some sequential processes, such as rolling or drawing, large reductions of the material thickness can be achieved. However, the shapes produced by these processes are not bulky enough to be used for further conversion into products. Thus, new metal forming processes capable of generating very large or severe plastic deformation (SPD) without major change in the billet geometry have been developed. SPD is a generic term describing a group of metal working techniques involving very large strains, typically involving a complex stress state or high shear, resulting in a high defect density and equiaxed "ultrafine" grain (UFG) size (d<500 nm) or nanocrystalline (NC) structure (d<100 nm). The metal working methods/techniques which can impose very large strains typically involving a complex stress state or high shear, resulting in a high defect density and equiaxed micro/nano-grain-sized structure, are generically called "Severe plastic deformation (SPD)" an exclusive technology in the top-down category. The encouraging economic forecasts for SPD cause great interest in the development of new fabrication techniques, with mass production scale-up capabilities and also at low cost.

This book focusses on techniques and the outstanding properties of materials with ultrafine/nano-grained structure, and also considers the basic understanding behind. Since only recently such materials have been produced in bulk shape opening their use in commercial and industrial applications. This book also describes critical challenges and commercial products fabricated by SPD methods.

A briefing about the classification of grains, fundamentals of plastic deformation, effect of plastic deformation on grain refinement, grain orientation and texture is given in the following sections for better understanding.

1.3 CLASSIFICATION OF GRAINS

The polycrystalline metals are identified as coarse grained, ultrafine-grained and nano-grained based on the grain size regimes. Grain sizes greater than 1000 nm are called as coarse-grained polycrystalline metals where traditional mechanisms determine deformation. In the range from 1000 nm down to 30 nm, called as ultrafine-grained polycrystalline metals where disordered grain boundaries begin to dominate the mechanical behaviour. This transition becomes much more evident below100 nm. Below 30 nm, the grains are termed as nano-grained polycrystalline metals where the atomic sliding at grain boundaries increases, leading to virtually no further work hardening of the plastically deformed metal.

One important method for grain refinement in metals is by deformation. Plastic working remains efficient in terms of producing small grain sizes only in the range down to 30 nm for most metals and their alloys. Employing metal forming, it is only possible to fabricate so-called ultrafine-grained (UFG) metals, not even touching desirable nanometals. However, new findings on the mechanism of plastic flow of nanometals could shed light and open the gate for metal forming also in this range.

1.4 BASIC OF PLASTIC DEFORMATION

It is a well-known fact that all solid materials can be deformed when subjected to external load. Up to certain limiting loads, a solid recovers its original dimensions when the load is removed and is known as elastic behaviour. Beyond which material no longer behaves elastically is the elastic limit. If the elastic limit is exceeded, the body experiences a permanent set or deformation when the load is removed and is said to have undergone plastic deformation. In case of elastic deformation, the atoms are stretched from their equilibrium position without breaking their chemical bonding with its neighbouring atoms. Hence, upon releasing the load, the atoms will be back to their initial equilibrium position, and the solid recovers its original dimensions. But in plastic deformation, atoms are displaced by a sufficient distance from their initial position that a new equilibrium location is achieved. Chemical bonds binding the atoms to their nearest neighbours are broken and reformed, hence upon releasing the load, the atoms will not come to their initial equilibrium position from the new equilibrium position, resulting in a permanent change of dimensions.

This ability to undergo plastic deformation is due to a combination of two factors: bonding and the crystal structure of metals. Of these, metallic bonding is notable because it allows the rapid and easy disruption and formation of bonds. Compared to covalent bonding, which involves the sharing of electrons in specific orbitals (and so is extremely dependent on both atom type and orientation), or ionic bonding, where cohesion is provided by strong electrostatic interactions between positive and negative ions (which are thus very sensitive to ion type and location), metallic bonding is mediated by a sea of delocalised "conduction" electrons that may freely travel through and around the lattice of positively charged metal ions. As such, bonding in metals is comparatively easily disrupted and reformed. The second factor that assists with the ductility of metals is their crystalline structure. The impact of this can be seen in the formation of amorphous metallic glasses that show a greatly decreased ductility despite a similar degree of metallic bonding. This is because the repeating structure of the crystal allows strain to be taken up by either the translation or transformation of the crystal.

The second factor is the crystalline structure. A crystal structure is made up of atoms. A crystal lattice is made up of points. A crystal system is a set of axes. In other words, the structure is an ordered array of atoms, ions or molecules. Based on their point groups, crystals and space groups are divided into seven crystal systems, namely, triclinic, monoclinic, orthorhombic, trigonal, tetragonal, hexagonal and cubic systems. Among these seven systems, most of the metals fall under the cubic system, and a few are in the hexagonal system. Further discussions are restricted only to cubic and hexagonal systems. Before discussing the basics of plastic deformation, a brief introduction about the basics of cubic and hexagonal crystal geometry and its planes and directions will be useful for better understanding.

1.4.1 Crystal Geometry

X-ray diffraction studies show that the atoms in a metal are arranged in a regular, repeated three-dimensional pattern. The atomic arrangements of metals are most simply portrayed by a crystal lattice in which atoms are visualised as hard balls located at particular locations in a geometrical arrangement. The most elementary crystal structure is the simple cubic lattice (Figure 1.1).

This is the type of structure found in ionic crystals such as NaCl and LiF, but not in any of the metals. Three mutually perpendicular axes are arbitrarily placed through one of the corners of the cell. Crystallographic planes and directions will be specified in terms of the length of their intercepts on the three axes, measured from the origin of the coordinate axes. To simplify the crystallographic formulas, the reciprocals of these intercepts are used. They are reduced to a lowest

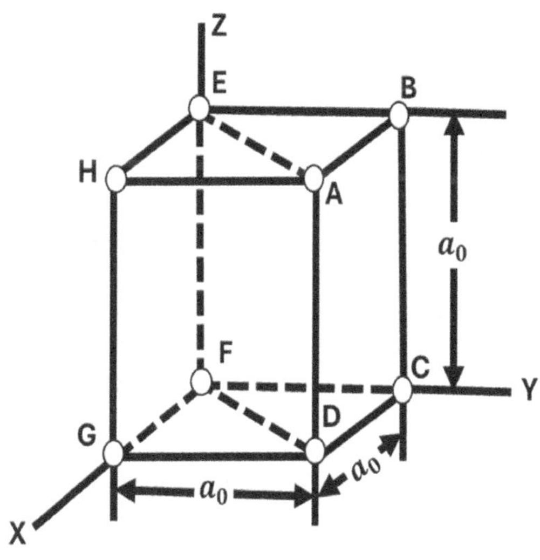

FIGURE 1.1 Simple cubic structure.

common denominator to give Miller indices (hkl) of the plane. For example, the plane ABCD in Figure 1.1 is parallel to the x and z axes and intersects the y axis at one interatomic distance (a_0). Therefore, the indices of the plane are $1/\infty$, $1/1$, $1/\infty$, or (hkl)=(010). Plane EBCF would be designated as the ($\bar{1}00$) plane, since the origin of the coordinate system can be moved to G because every point in a space lattice has the same arrangements of points as every other point. The bar over one of the integers indicates that the plane intersects one of the axes in a negative direction. There are six crystallographically equivalent planes of the type (100), any one of which can have (100), (010), (001), ($\bar{1}00$), ($0\bar{1}0$), ($00\bar{1}$) depending upon the choice of axes. The notation {100} is used when they are to be considered as a group, or family of planes. Crystallographic directions are indicated by integers in brackets: [uvw]. Reciprocals are not used in determining directions. As an example, the direction of the line FD is obtained by moving out from the origin a distance a_0 along the x-axis and moving an equal distance in the positive y direction. The indices of this direction are then [110]. A family of crystallographic equivalent directions would be designed <uvw>. For a cubic lattice only, a direction is always perpendicular to the plane having the same indices.

In order to specify planes and directions in the hexagonal close-packed (hcp) structure, it is convenient to use the Miller-Bravais system with four indices of the type (hkil). These indices are based on four axes: a_1, a_2 and a_3 are 120° apart in the basal plane, and the c axis is normal to the basal plane. These axes and typical planes in the hcp crystal structure are given in Figure 1.2. The third index is related to the first two by the relation $i=-(h+k)$.

1.4.2 Deformation Mechanisms

In many metals, the basic mechanism of plastic deformation is slip and twinning. However, in cases where slip is not possible, twinning becomes the basis of plastic deformation.

The first of these mechanisms, the translation of planes relative to one another, is known as *slip*. In slip, a plane of atoms is translated one full position across; this retains the crystal structure of the material, but leaves it displaced along a clear boundary known as a *slip plane*, which occurs preferentially along planes of higher atomic density and the slip direction within the slip plane. Since the planes of greatest atomic density are also most widely planes in the crystal structure, the resistance to slip is generally less for these planes than for other set of planes. The slip plane, along with the slip direction, is called as slip system.

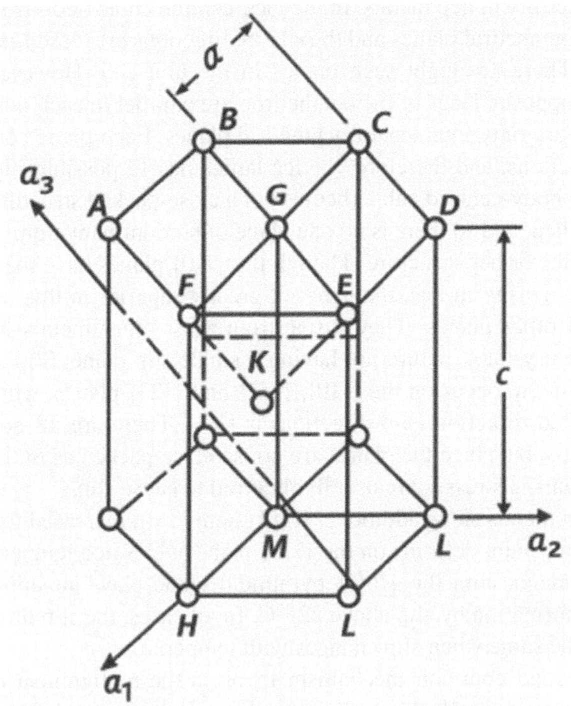

Basal plane (0001) – *ABCDEF*
Prism plane* (1010) - *FEJH
Pyramidal planes
Type I, Order 1 (1011) - *GHJ*
Type I, Order 2 (1012) - *KJH*
Type II, Order 1 (1121) - *GHL*
Type II, Order 1 (1122) - *KHL*
Diagonal axis [1120] - *FGC*

FIGURE 1.2 Hexagonal close-packed structure.

In the hcp metals, the only plane with high atomic density is the basal plane (0001). The axes <1120> are the close-packed directions. Since there is only one basal plane per unit cell and three <1120> directions, the hcp structure possesses three slip systems being the reason for low ductility in hcp metals. In the face-centred cubic (fcc) structure, the {111} octahedral planes and the <110> directions are closed-packed systems. There are eight such planes in the unit cell. However, the planes at opposite faces of the octahedron are parallel to each other, so that there are only four sets of octahedral planes. Each plane contains three directions, and therefore, the fcc lattice has 12 possible slip systems. The body-centred cubic (bcc) is not a close-packed structure like the fcc or hcp, and so there is no one plane of predominant atomic density as in fcc or hcp structure. Though the {110} planes have the highest atomic density in bcc, they are not greatly superior in this respect to several other planes. They differ from most other metals having fcc or hcp crystal structure not having a single slip plane. Slip in bcc metals found to occur on the {110}, {112} and {123} planes, while the close-packed direction (slip direction) is [111]. There are 48 possible slip systems, but since the planes are not so close-packed as in the fcc, higher shearing stresses are usually required to cause slip.

Certain metals show additional slip systems with increased temperature. Aluminium deforms on the {110} plane at elevated temperature, while in magnesium the $\{10\bar{1}1\}$ pyramidal plane plays an important role in deformation by slip above 225°C. In all cases, the slip direction remains the same when slip changes with temperature.

The second common mechanism involves the realignment of the crystal structure itself; this may occur through the direct deformation of the crystal due to shear stresses or may be mediated by transformation into an intermediate phase that then reverts to a different orientation of the parent phase. In either case, the result is a reoriented crystal structure that is *twinned* along a specific *twinning plane*. Slip and twin-mediated deformation processes are illustrated schematically in Figure 1.3.

It should be noted that twinning differs from slip in several respects. In slip, the orientation of the crystal above and below the slip plane is the same after deformation as before, while twinning results in an orientation difference across the twin plane. Slip is usually considered to occur in discrete multiples of the atomic spacing, while in twinning, the atom movements are much less than an atomic distance. Slip occurs on widely spread planes, while in the twinned region of a crystal, every atomic plane is involved in the deformation.

Twins may be produced by mechanical deformation or as a result of annealing following plastic deformation. The first type is known as mechanical twins; the latter are called annealing twins. Mechanical twins are produced in bcc or hcp metals under conditions of rapid rate of loading (shocking loading) and decreased temperature.

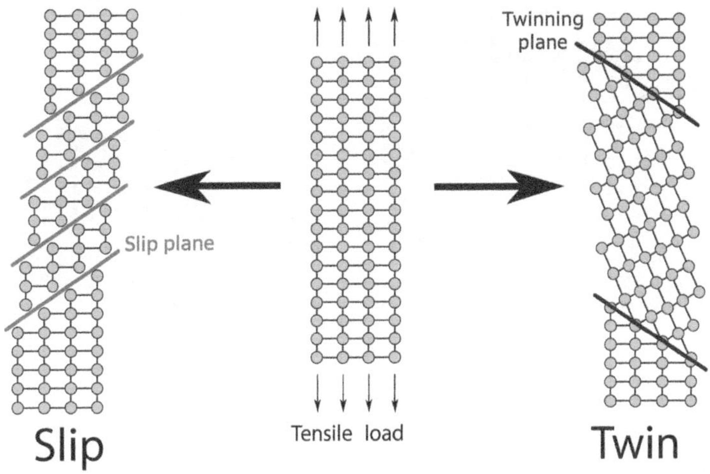

FIGURE 1.3 Common forms of plastic deformation – slip and twin.

FCC metals are not ordinarily considered to deform by mechanical twinning, although Au–Ag alloys twin fairly readily when deformed at low temperature and mechanical twins have been produced in copper by tensile deformation at 4 K and by shock loading. Twins can form in a time as short as a few microseconds, while for slip, there is a delay time of several milliseconds before a slip band is formed. Under certain conditions, twins can be headed to form with a click or loud noise (tin cry). Like slip, twinning occurs in a definite direction on a crystallographic plane and direction for each crystal structure. Common twin planes and directions in bcc, hcp and fcc are (112), [111]; ($10\bar{1}2$), [$\bar{1}011$] and (111), [112], respectively. Twinning occurs generally when the slip systems are restricted or when the critical resolved shear stress increases, and it is not a dominant deformation mechanism.

In both cases, this deformation is fundamentally associated with the motion of atoms, but the nature of this atomic motion differs. For slip, the deformation occurs on an individual atom scale, driven by the motion of defects in the crystal structure through the material. These defects, known as *dislocations*, occur most frequently as *edge* and *screw* dislocations; edge dislocations can be thought of as an extra plane of atoms that terminates halfway through the crystal lattice, while screw dislocations involve a shearing of the lattice such that the crystal planes are linked in a helical pattern around the dislocation. Edge and screw dislocations are illustrated schematically in Figure 1.4.

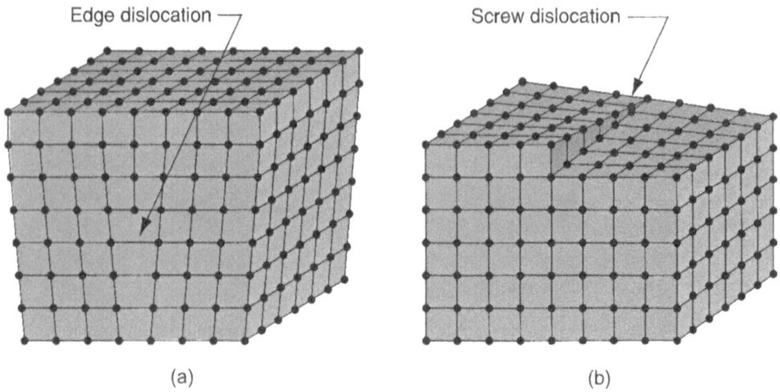

FIGURE 1.4 Crystal dislocations (a) edge and (b) screw.

A dislocation is the linear lattice defect responsible for nearly all aspects of the plastic deformation of metals. As a dislocation moves through the material, atoms are translated in individual steps, and eventually form a dislocation step on the surface of the grain. As dislocations are formed through applied stress, transmitted through the crystal, and exhausted on the grain surface, these translations build up, resulting in slip and eventually producing macroscale deformation. A consequence of the dependence of slip-on dislocation motion is that inhibiting the motion of dislocations will hinder plastic deformation; introducing grain boundaries, secondary phases, solute atoms of different dimensions, or even additional dislocations can all hinder dislocation motion. These processes, known respectively as grain refinement, precipitation strengthening, solid-solution strengthening, and work hardening, can thus all increase the yield strength of the material, albeit typically at the cost of decreased ductility.

While slip occurs through the discrete motion of full edge and screw dislocations, twinning is more diverse and can proceed through a number of mechanisms. Twinning may also proceed through a coordinated process, with multiple atoms moving synchronously; this can lead to a very rapid growth of the twin, proceeding through the material at the speed of sound. Indeed, because these happen so fast, sufficient volumes of the material may be deformed to produce an audible sound. In twinning, each dislocation represents a smaller displacement than the "perfect" edge and screw dislocations.

Slip and twinning occur to some extent in all metallic systems, but which mechanism dominates is dependent on the crystal structure and the environment. Slip is favoured along close-packed crystal planes, as these represent smaller displacements of individual atoms, and correspondingly lower energy barriers. Certain *slip directions* will also be favoured, with motion of the dislocations in this direction requiring the crossing of a smaller energy barrier; the combination of a slip direction and slip plane is known as a *slip system*. Slip occurs when the component of the applied stresses in a given slip plane and direction is sufficient to drive the dislocation through the energy barrier, for example, when they exceed the *critical resolved shear stress* (τ_{CRSS}), given as

$$\tau_{CRSS} = \sigma \cos(\phi) \cos(\lambda)$$

with σ as the applied stress, ϕ as the angle between σ and the slip plane, and λ as the angle between σ and the slip direction.

Due to both the requirements for close-packed planes and preferentially aligned slip systems, slip is therefore favoured in crystal systems with multiple distinct slip systems. This is clearly seen in the fcc lattice, which has numerous distinct close-packed planes and is typically highly ductile. Conversely, in the hexagonal close-packed (HCP) lattice, there is only one close-packed plane, and so slip in other directions is difficult; here, deformation by twinning is favoured. These limitations can have a profound effect on ductility; in particular, materials that do not satisfy the *von Mises criterion* of at least five independent slip systems are unable to sustain significant plastic deformation by slip, and so show limited ductility. Despite this, substantial slip may still be noted in even HCP lattices, so long as sufficient thermal energy is provided to overcome the energy barrier to dislocation motion; this can be seen when heating metals, which allows them to become more ductile and malleable. The reverse of this is also seen, as cooling metals may cause certain slip planes to become deactivated. Through these mechanisms, substantial deformation of the material is possible; elastic strains are typically limited to only 1%–2%, while plastic strains in ductile materials may exceed even 50%.

1.4.3 Plastic Deformation and Grain Refinement

1.4.3.1 Plastic Deformation

At room temperature, the yield stress of metallic materials increases with decreasing grain size. This is known as the Hall-Petch relationship. One of the possible techniques used to obtain small grains in metals consists of applying a large level of plastic strain to a coarse-grain precursor. A simultaneous accumulation of localised dislocations and an increase in the lattice misorientation are responsible for crystal subdivision and subsequently develop sub-microscopic grains. Besides, dynamic recrystallisation due to the plastic deformation also produces grain refinement as a resulting microstructure in a few cases. Refining grains of metals and alloys is an effective approach to tailor their properties and performance. Plastic deformation is routinely used for structure refinement, with which microstructures down to the submicron scale can be accessed in most metals and alloys.

On applying the stress, the dislocations move, causing plastic deformation by slip or twinning. The movement of dislocations is hindered by grain boundaries, second phase particles and even with sessile dislocations, i.e., dislocations which cannot move. The mobility is based on the characteristic features, i.e., Burgers vector of dislocations. The Burgers vector (b) is the vector that defines the magnitude and direction of slip. In pure edge dislocation, the Burgers vector is perpendicular to the dislocation line, while for screw dislocation, the Burgers vector is parallel to the dislocation line. The macroscopic slip produced by edge and screw dislocation is identical, but for an edge dislocation line moves parallel to the slip dirction while the screw dislocation moves at right angles to it. In real conditions, dislocations are rarely lie in a single plane or pure. Dislocations will be partly edge and partly screw in character. Dislocations will take the form of curves or loops, which in three dimensions form an interlocking dislocation network. However, in a slip plane, the dislocation line can be resolved into an edge or screw component.

Dislocations can slip in planes containing both the dislocation line and the Burgers vector, the so-called glide plane. For a screw dislocation, the dislocation line and the Burgers vector are parallel, so the dislocation may slip in any other slip plane, which is called as cross slip. However, for an edge dislocation, the dislocation and the Burgers vector are perpendicular, so there is one plane in which the dislocation can slip, and it can not move to another slip plane, i.e., cross slip is not possible if the component of the dislocation loop has edge characteristics. An alternative mechanism of dislocation motion that allows an edge dislocation to move out of its slip plane is climb. The driving force for dislocation climb is the movement of vacancies through a crystal lattice. If a vacancy moves next to the boundary of the extra half plane of atoms that forms an edge dislocation, the atom in the half plane closest to the vacancy can jump and fill the vacancy. Climb is temperature-dependent. Climb occurs much more rapidly at high temperatures than at low temperatures due to an increase in vacancy motion.

The dislocations moving in its slip plane will intersect other dislocations crossing the slip plane. The intersection of dislocations plays an important role in strain hardening and in the formation of sub-grains. The intersection of two dislocations produces a sharp break, a few atom spacing in length, in the dislocation line; these breaks can be of two types: (i) Jog – a sharp break in the dislocation moving it out of the slip plane and (ii) kink – a sharp break in the dislocation line but remains in the slip plane. Basically there are four types of intersections; (i) intersection of two edge dislocations with Burgers vector at right angles to each other produces a jog with edge orientation and therefore it can glide with the rest of the dislocations but can not cross slip, (ii) intersection of two edge dislocations with parallel Burgers vector produces jog in both dislocations with screw orientation and lie in the original slip plane. Jogs of this type are called as kinks. They are unstable since during glide they can line up and annihilate the offset, (iii) intersection of a screw and edge dislocation produces a jog with an edge orientation on edge dislocation and a kink with edge orientation on screw dislocation and therefore screw dislocation can not cross slip, (iv) intersection of two screw dislocations produces jogs of edge orientation in both screw dislocation. Both the screw dislocations can glide but cannot cross slip. They can move only by non-conservative motion called climb, which is a thermally activated process. This is the most important type of intersection from the viewpoint of plastic deformation.

Dislocations frequently pile up on slip planes at barriers such as grain boundaries, second phases or sessile dislocations produced due to the interaction of dislocations as discussed previously. Pile up of dislocations leads to a high concentration of stress on the leading dislocations in the pile up which can approach the theoretical shear stress of the crystal. This high stress can either initiate yielding or it can nucleate a crack at the barrier.

1.4.3.2 Grain Boundary and Grain Refinement

The grain boundaries in a polycrystalline aggregate in a region of disturbed lattice are only a few atomic diameters wide. The crystallographic orientation changes abruptly between the grains. If the difference in crystallographic orientation between adjacent grains is more than 15°, it is considered as high angle grain boundaries (HAGB) and if the difference is less than 15° it is considered as low-angle grain boundaries (LAGB). However, the misfit angle is arbitrary. The unorganised structure of atoms belongs to both grains, while most belong to neither. Essentially, they are dislocations called as grain boundary dislocations. These are not mobile dislocations. A low-angle boundary contains relatively simple arrangments of dislocations. Along the boundary, the atoms adjust their position by localised deformation to

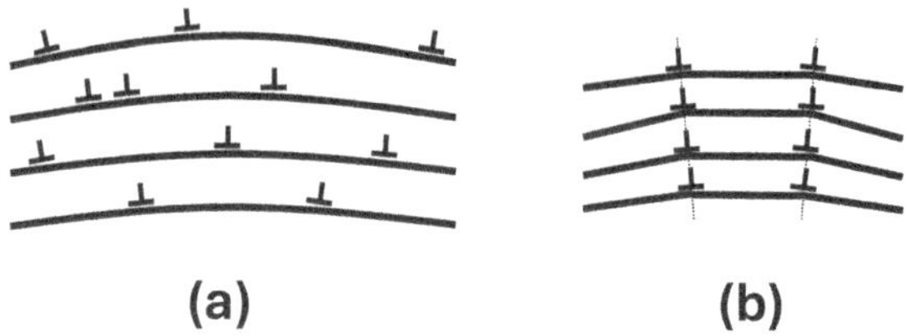

FIGURE 1.5 Movement of dislocations to produce polygonisation.

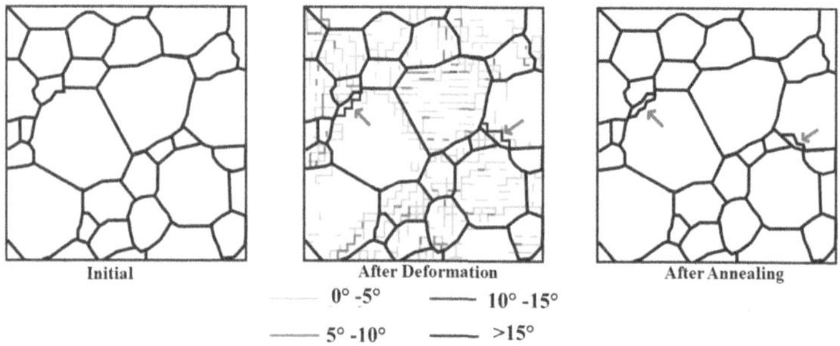

FIGURE 1.6 Formation of sub-grains on deformation and annealing.

produce a smooth transition from one grain to the other. Where the atom planes end, there is an edge dislocation. This process has been called insitu recrystallisation or polygonisation, which was used originally to describe the situation that occurs when a single crystal is bent to a relatively small radius of curvature and then annealed. Bending results in the introduction of an excess number of dislocations of one sign. The dislocations are distributed along the bent glide planes as shown in Figure 1.5a. When the crystal is heated, the dislocations group themselves into the lower energy configuration of a low-angle boundary by dislocation climb. The resulting structure is a polygonlike network of LAGB (Figure 1.5b).

In polycrystalline materials, low-angle boundaries, also called as sub-boundaries, can be produced in a number of ways. They may be produced during crystal growth, during high temperature deformation or as a result of phase transformation. Perhaps the most general method of producing is by introducing deformation and followed with annealing to rearrange the dislocations into subgrain boundaries (Figure 1.6). The amount of deformation and temperature must be low enough to prevent the formation of new grains by recrystallisation.

The amount of subgrain boundary formation is higher when imposed strain (deformation) is higher, and hence in severe plastic deformation, grain size in the range of ultrafine and/or nano size can be obtained easily. Though the material after (severe) deformation achieves higher strength due to fine grains, the ductility will be poor due to high dislocation density, which prevents the movement. However, after annealing, due to polygonisation, the ductility can be improved, accepting a small sacrifice in strength. By suitably choosing the reduction amount, annealing temperature and time, material with higher strength and comparable ductility as compared to the material which is in annealed condition can be obtained.

2 Severe Plastic Deformation Processes Developed for Grain Refinement

2.1 INTRODUCTION

SPD processes, originally proposed for grain refinement, along with equivalent strain is shown in Figure 2.1.

High-pressure torsion (HPT) refers to the processing of metals whereby samples are subjected to a compressive force and concurrent torsional straining. In HPT, a small disc sample or small amount of powder is squeezed between two anvils under high pressure and concurrently strained by rotating the lower anvil with respect to the

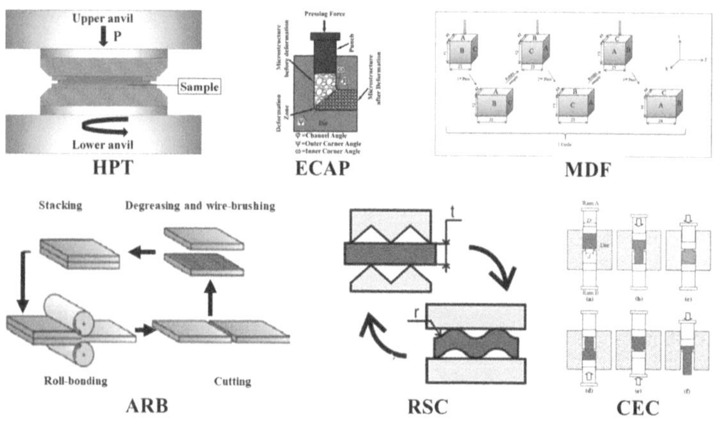

FIGURE 2.1 SPD processes employed for grain refinement.

DOI: 10.1201/9781003596110-2

upper anvil. Equal channel angular pressing (ECAP) is a processing method in which a metal is subjected to intense plastic straining through simple shear without any corresponding change in the cross-sectional dimensions of the sample since the dimensions at the entry and exit channels are the same. The complex deformation of the metal as it flows around the corner produces very high strain. Because the cross section remains the same, a workpiece can be extruded multiple times with each pass introducing additional strain. In multi-directional forging, each sample is initially forged in one direction, and then, it is rotated by 90° and forged in that direction, constituting a pass. Alternate forging of these alloys takes place in order to achieve bulk grain refinement without any accompanying change in the cross-sectional dimensions of the sample. By subjecting the material multiple times, high strain can be imparted to the sample. This leads to the accumulation of plastic strain in the material. In the accumulative roll bonding (ARB), stack of metal sheets, which are repeatedly rolled to a severe reduction ratio, sectioned into two halves, piled again and rolled. Thus, the strain gets accumulated with strain leading to grain refinement. Repetitive corrugation and straightening (RCS) involves the repeated bending and straightening of a workpiece to create fine-grained structures in metallic sheets or plates without a change in the cross-section geometry of the workpiece. Due to repeated bending and straightening, a large plastic strain is introduced leading to grain refinement. It induces more homogeneous microstructure in the metal and mechanical properties will be influenced by the grain refinement. During the CEC deformation process, the intense plastic strain can be achieved by pressing the sample between the extrusion and compression chambers alternately as there is no change in shape and size of the sample before and after the process. An advantage of the CEC process is that the material can undergo several cycles without being removed from the container reducing material handling and losses.

The way of calculating plastic strain, the main process parameter, has been different for different processes (Table 2.1). Plastic strain is a continuous relationship in HPT. However, in most cases, the required strain will be accumulated in a recurring way, depending on a single pass, on tool characteristics (ECAP, CEC), on billet dimensions (MF, ARB) or a combination of both (RCS). The details are discussed in subsequent sections.

TABLE 2.1
Plastic Strain in Important SPD Processes

Process	Plastic Strain (ε)
HPT	Shear strain, $\gamma = r\theta/h$, Where, r is the distance from the centre of disc, $\theta = 2\pi N$ is the rotation angle in radian (N is the number of turns) and h is the thickness of disc. Linear strain, $\varepsilon = \gamma/\sqrt{3}$
ECAP	$\varepsilon_N = \frac{N}{\sqrt{3}}\left[2\cot\left(\frac{\Phi}{2}+\frac{\Psi}{2}\right)+\Psi\,\mathrm{cosec}\left(\frac{\Phi}{2}+\frac{\Psi}{2}\right)\right]$ where N – No. of passes, ϕ is channel angle, ψ is Outer corner angle.
MDF	$\varepsilon_{eff} = N\,(2/\sqrt{3})\,\ln\,(h_0/h)$ where N is the number of passes carried out in the sample, which needs three passes to complete one cycle of MAF. "h_0" and "h" are initial and final heights of the samples.
ARB	$\varepsilon_{eff} = N\,(2/\sqrt{3})\,\ln\,(t_0/t)$ where N is the number of passes carried out in the sample, "t_0" and "t" are the initial and final thickness of the samples.
RCS	$\gamma_{xy} = \Delta x/t = \tan\theta$ and $\varepsilon_{xy} = \gamma_{xy}/2$ where γ_{xy} is the shear strain, Δx is the sheared distance, through an angle of θ, and t is the thickness of the sheet sample.
CEC	$\varepsilon = 4N\,\ln\,(D/d)$ where N is the number of passes, D and d are the die channel diameter in the outer and the small middle region, respectively.

2.2 RECENT DEVELOPMENTS

In recent years, new developments based on the existing SPD techniques have emerged, which can be split into two categories: (i) redevelopment of originally invented SPD techniques or equipment of existing metal-forming processes with intense shearing and (ii) inventive use of rotational tool movement, essential for performing uninterrupted shear straining.

Figure 2.2 shows a few rotating tool configurations that add new features into the grain refinement process and reduce strain gradient in the radial direction; Processing long products in a cyclic twist forward extrusion; friction stir welding modified into a continuous grain refining process. From the customer's point of view, categorisation based on the form of the product, i.e., slug, rod, wire, hoop, band and sheet, is of some importance. Most techniques enable the processing of a finite portion of metal. Some of them, derived from well-established metal-forming methods, work with narrow continuous billets.

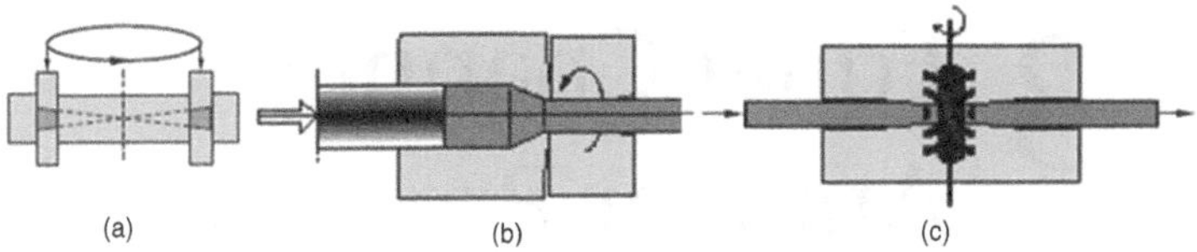

FIGURE 2.2 Different applications of twisting in SPD processes: (a) HPT of rings, (b) cyclic twist extrusion and (c) friction stir welding.

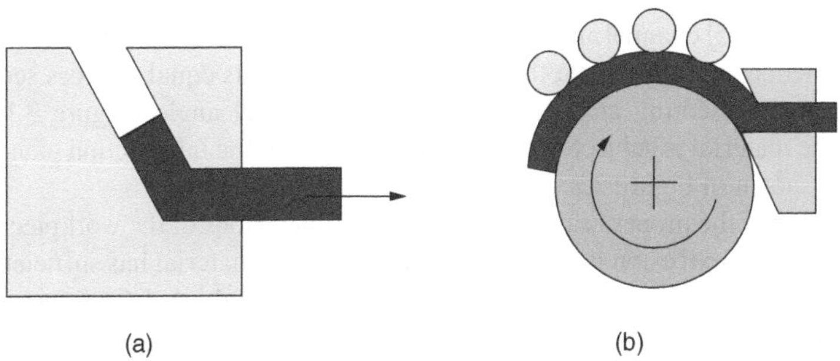

FIGURE 2.3 Continuous SPD processes derived from metal-forming technology: (a) equal channel angular drawing ECAD and (b) continuous confined strip shearing – CONSHEARING.

ARB is the first SPD process that dealt with the material in the sheet form while retaining the sheet thickness. RCS is capable of producing UFG rings. Developing SPD methods towards the production of tailored products made of UFG metals would be relatively easy. However, it has not been tried since the market for nanometals remains in an embryonic phase. The key to the wide commercialisation of UFG materials is to lower their processing cost and waste. It is believed that it can be achieved through continuous processing. Continuous SPD processes, based on popular metal-forming operations, e.g., wire drawing and continuous extrusion, are illustrated in Figure 2.3.

One of the challenges is that the many pre-processing (i.e. grain refining) and post-processing (i.e. shaping) techniques are reaching their limits for applications involving high-strength nanostructured materials. At present, the fundamental SPD pre-processes shown in Figure 2.1 are still mostly employed for grain refinement; meanwhile, they are being improved and made robust for industrial use. The recent developments in each of the SPD processes are discussed in detail while discussing the processes individually in the respective chapters.

3 Equal Channel Angular Pressing

3.1 PRINCIPLE AND PROCESS

In the equal channel angular pressing (ECAP) process, a well-lubricated billet is pressed in a die that contains two channels equal in cross section, intersecting at an angle called die channel angle (Figure 3.1). The material is subjected to shear deformation at the intersection plane, as shown in Figure 3.2.

Since the process can maintain the net dimensions of the workpiece, repetitive extrusion is possible, provided that the material has sufficient ductility to withstand heavy deformations. The multiple ECAP passes make it possible to rotate the billet around its longitudinal axis between the passes, creating different routes.

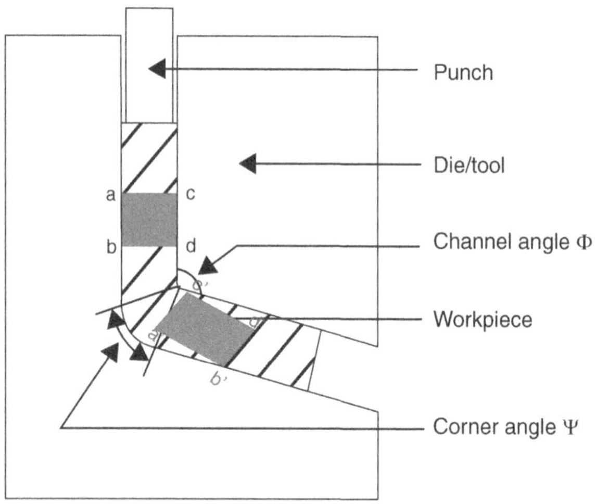

FIGURE 3.1 Schematic view of ECAP die.

DOI: 10.1201/9781003596110-3

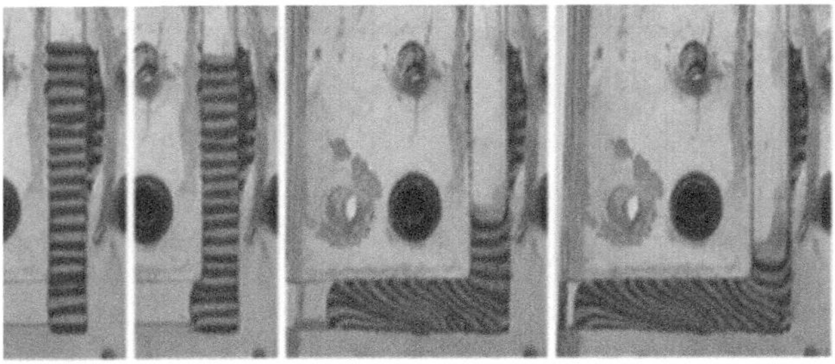

FIGURE 3.2 Shearing pattern in the intersection angle.

3.2 SHEAR STRAIN AND EQUIVALENT STRAIN

The improvement in strength is due to the strain hardening of the material. The amount of strain induced in turn depends on the die channel angle and die corner angle. In ECAP, metal flow through the channels and subjected to simple shear and characterised by several fundamental parameters such as the strain imposed in each separate passage through the die, the slip systems operating during the pressing operation and the consequent shearing patterns present within the as-pressed billets. Though all of these parameters play a critical role in determining the nature of the ultra-fine grained (UFG) structure introduced by ECAP, the most important is strain per pass.

An abrupt strain is imposed on a sample in every passage through an ECAP die. The magnitude of this strain is based on the die configurations such as channel angle (ϕ) and angle associated with the arc of curvature where the two parts of the channel intersect, known as corner angle (ψ) as shown in Figure 3.1. In strain calculation, assuming a fully lubricated specimen, the frictional effects are neglected. A small square element in the entrance channel, labelled "abcd", passes through the theoretical shear plane and becomes distorted into the parallelogram labelled a′b′c′d′ as shown in Figure 3.1. The shear strain (γ) can be calculated from first principles.

When $\psi = 0°$, $\gamma = 2 \cot(\phi/2)$

When $\psi = (\pi - \phi)°$, $\gamma = \psi$

When $\psi = 0° < \psi < (\pi - \phi)°$,

$$\gamma = 2\cot\left(\frac{\Phi}{2}+\frac{\Psi}{2}\right)+\Psi\,\mathrm{cosec}\left(\frac{\Phi}{2}+\frac{\Psi}{2}\right)$$

Using von Mises criterion, the equivalent strain (ε) can be expressed as ($N\,\gamma/\sqrt{3}$) where N is the number of passes.

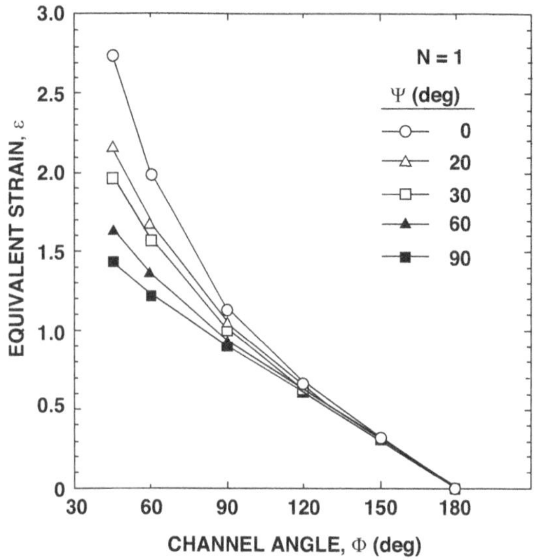

FIGURE 3.3 Graphical representation of equivalent strain with channel and corner angles.

The graphical representation of equivalent strain and channel angle for different corner angles for single pass is shown in Figure 3.3.

Many conclusions can be made from Figure 3.3. The most important thing is that the corner angle has a relatively minor effect on the equivalent strain when the channel angle is more than 90°. Additionally, high strains can be achieved in a single pass by constructing a die with very low values of ϕ and ψ. The split dies with acute angles are practically difficult since the force (stress) required for ECAP will be enormous, which will create a problem in the design of grips or the bolts. Conventionally dies are designed to have a channel angle of 90° so that the equivalent strain is close to 1 for a single pass, and this strain is essentially independent of the corner angle. The equivalent strain (ε) can be expressed as

$$\varepsilon = \left(2N/\sqrt{3}\right)\cot\left(\phi/2\right)$$

3.3 INFLUENCE OF EXTERNAL CONDITIONS

3.3.1 Effect of Processing Route

Conventionally, ECAP has been classified into four routes based on the rotation of the billet between each pass with reference to its longitudinal

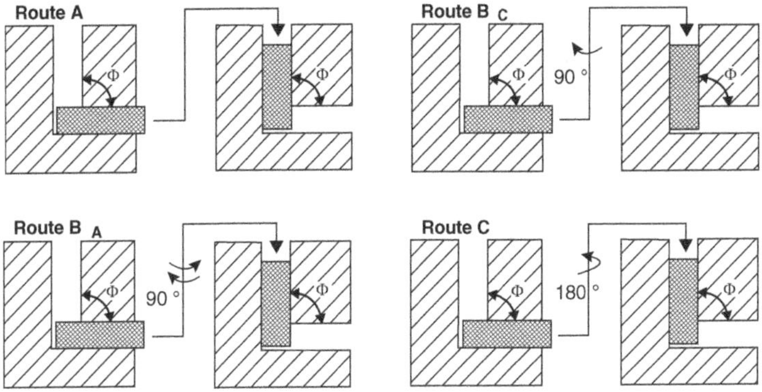

FIGURE 3.4 Rotation schemes of four ECAP routes.

axis as (i) Route A, no rotation (ii) Route B_A, rotating 90° in opposite directions, (iii) Route B_c, rotating 90° in the same direction and (iv) Route C, rotating 180°. Figure 3.4 schematically illustrates the four processing routes of ECAP.

The shear planes in the extrusion axis (X direction) and the perpendicular axes (Y and Z direction) tend to rotate based on the channel angle and processing route (Figure 3.5). The rotation causes changes in the shearing planes, resulting in (i) development of slip traces in the three planes within the sample, (ii) regular and periodic rotation of equiaxed structure and (iii) the occurrence of deformation in all three planes. These conditions are readily satisfied by Route B_c. Hence, the grain refinement and properties vary with respect to different routes.

In Route A, the amount of deformation of the billet surface keeps reducing as the number of passes increases because the deformation in the billet ends is very little in subsequent passes. On the other hand, in Route C, the billet ends recovered their undeformed shape at each even number of passes and the strain in middle of the billet accumulated in each pass, and hence Route C develops a more stable deformation zone. In Route A, the grain shape was highly directional. In Route B, grains were equiaxed, and in Route C, the grains were most equiaxed and coarser since in Routes B and C, the strain distribution was more uniform, while it was more inhomogeneous in Route A and monotonous in one direction. Effectiveness of a route is also based on the crystal structure, slip systems and stacking fault energy. For example, Cu and Al deform by slip, Ag prefers twinning since the stacking fault energy of Ag is lower than for Cu or Al. The HCP metals, due to a lack of slip systems, prefer twinning, which is based on the c/a ratio. So, the

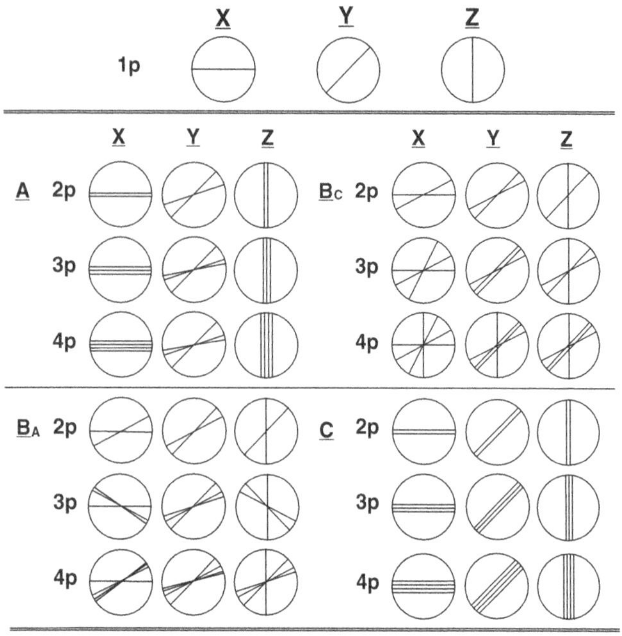

FIGURE 3.5 Shearing pattern on X, Y and Z planes for the processing routes.

reversal of strain and redundant strain during processing may not be effective for HCP metals.

3.3.2 Channel Angle and Corner Angle

Microstructural and continuum modelling suggest that the corner angle has a greater role as compared to the die angle with respect to strain rate sensitivity due to the widening of the localised zone in the die. Metals like titanium are greatly influenced by the strain rate, especially at warm temperatures. In most metals, better distribution of effective strain is achieved when the die corner angle is between 0° and 28° for a die channel angle of 90°. High strain homogeneity (ratio between strain in the investigated area to average strain in the entire billet) is obtained in the die with a channel angle of 90° and corner angle of 35° for a perfect plastic material when the friction coefficient is 0.3 in a single pass. However, this condition generates an effective strain of 0.66 only. Decreasing the corner angle to 0° and the friction coefficient to 0 generates a strain of 1.12, but strain homogeneity decreases to approximately 0.5 in single pass.

The deformation is more inhomogeneous in acute angles (less than 90°). Deformation is more complicated in acute angles compared to obtuse angles. The shape of the front end is more uniform in die with a channel angle of 90° and hence material wastage of the sample for multi-pass is less, leading to a greater steady state region in the sample. Effective strain contours across the width are more uniform in a 90° angle die as compared to any other angles.

As the channel angle (ϕ) increases, the effective strain decreases, and as the number of passes (n) increases and the effective strain increases. The inclination angle (β) of the elongated grains with respect to the extrusion axis is calculated as follows:

$$\beta = \arctan\left(\frac{\tan\phi}{2n}\right)$$

Inclination angle becomes smaller with a small channel angle and corner angle, and with larger frictional forces. The inclination angle between the extrusion axis and grain elongation becomes smaller with increasing number of passes in Route A as compared to other routes. Though the pressing efficiency (tendency towards inclination of grains towards pressing axis) is more in smaller die angles, the stresses on the dies will be more and designing will be very difficult, and moreover, the strain inhomogeneity will be more across the cross section. It is better to extrude with an optimum die angle, and efficiency can be increased by increasing the number of passes. From the above facts, the optimum channel angle is 90° with the corner angle between 0 and 35°.

3.3.3 Friction

Friction plays a major role in any forming operation, especially in determining the deformation load. The load due to friction varies with the type and form of lubricants used. The load due to friction is twice if solid MoS_2 is used instead of spray MoS_2. The high frictional forces in the solid lubricated die may be due to improper spreading of the lubricant, which promotes sticking friction conditions instead of hydrodynamic conditions.

Due to friction, gaps at four regions in the billet of elastic-plastic materials after a single pass are formed as shown in Figure 3.6. Due to shear deformation, the billet head tends to bend towards the top of the exit channel. It leads to the formation of gaps 1 and 3. The contact surface between die channel and side surface of head region forms gap 2

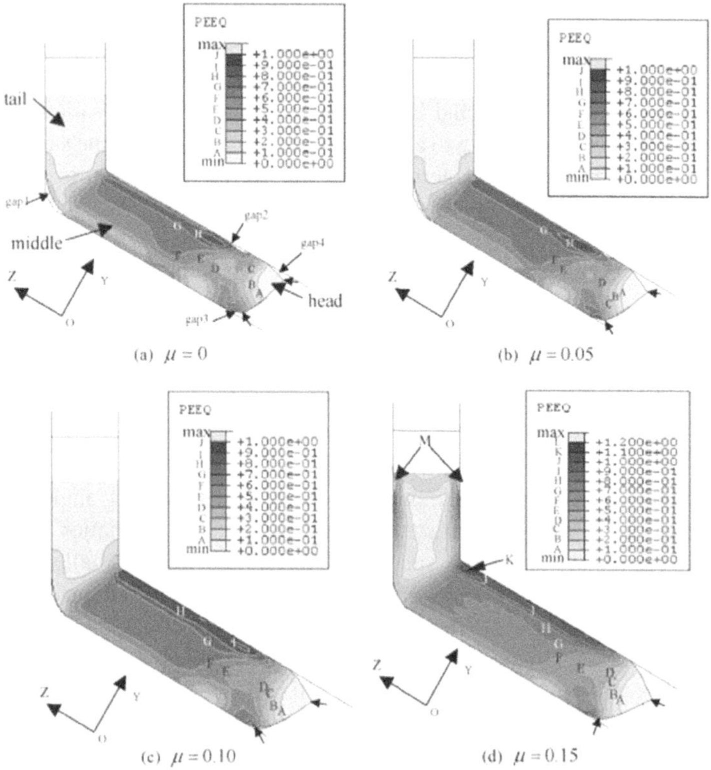

FIGURE 3.6 Gap formed in single pass ECAP of elastic-plastic material at different friction conditions.

and 4. The gaps 1, 2 and 3 reduce as the friction increases, but the reverse trend is observed for gap 4 as shown in Figure 3.6.

The effect of friction is very complex in three-dimensional conditions. The inhomogeneity factor in the Z plane increases, and in the Y plane it decreases, and in the X plane it is more complex as the friction coefficient increases. However, the distribution of strain is not affected much by friction as compared to the corner or die channel angle. The friction effect is very critical if the sample size is less than 5 mm, and hence, the minimum size should be more than 5 mm. However, tubes that normally lower wall thickness can be ECAPed with less pressure than a solid rod.

This is due to the following two reasons. In general, the higher the shape difficulty factor (ratio between surface area and volume), the higher the forming pressure. Interestingly, in ECAP of tubes with a moving mandrel, the extrusion pressure for the tube is lower by a factor of two than solid rod, though the shape difficulty factor for the tube is more (0.8) than solid rod (0.16). The sand mandrel moves along with the tube; the drag friction is in the same direction as the punch force as in the extrusion of tubes with moving mandrels. The friction between the mandrel and the tube is opposite to the friction between the tube and the die. So, the total force due to friction reduces, which is reflected in the total force reduction. The second reason is discussed in Section 3.4.2. Due to the opposite nature of friction in the inner and outer walls of the tube, even thin-walled tubes could be ECAPed.

3.4 INFLUENCE OF INTERNAL CONDITIONS

3.4.1 Influence of Strain Hardening and Strain Rate

Non-strain hardening material has a lesser shear zone as compared to strain hardening material. The deformation geometry in strain hardening materials is almost independent of die corner angle. However, the strain hardening rate has a negligible role in affecting effective strain. The deformation of strain hardening material is distinctly different from the flow pattern of the ideal materials. The leading billet end tends to separate from the die surface, and the billet cannot completely fill the die cavity. A gap between the billet and die corner appears, and hence the bottom side experiences less deformation as compared to the middle. The formation of the corner gap is due to the strain hardening behaviour. In addition to die corner gap, an exit channel gap is also formed between the billet top and the exit channel, as shown in Figure 3.7. The exit channel gap appeared in rate-sensitive material such as titanium during frictionless condition. The die corner gap not only reduces the overall shear but also results in strain inhomogeneity. It was also observed that the corner gap reduces with increasing the arc radius while the exit gap minimises when extruded at high strain rates.

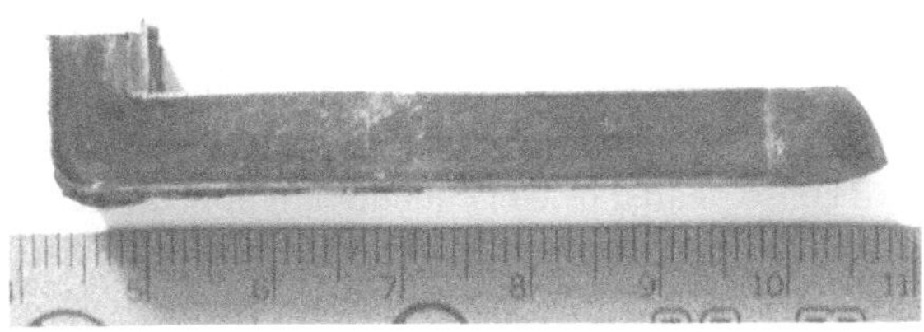

FIGURE 3.7 Die corner and exit corner gap in CP–Ti extruded at 300°C.

Increasing the strain rate during ECAP may result in higher dislocation density and finer grains, since the dynamic recovery is time-dependent. In general, at higher strain rates, the workability of materials is reduced. It is well known that materials are very sensitive to strain rates at high temperatures. The effect of strain rate is an important parameter during ECAP at warm temperatures.

3.4.2 Temperature

A temperature rise of about 15°C due to plastic deformation during each pass is observed. This temperature rise due to plastic deformation will help in reducing the flow stress locally. In the case of solid materials, the heat produced will be easily dissipated over a large area, and hence, a notable change in the punch force could not be realised. Interestingly, in ECAP of tubes with a sand mandrel, the adiabatic heating and its effect on extrusion pressure is observed clearly. The expected high pressure required for the tube due to the higher shape difficulty factor is reduced due to the heat produced during each pass. The retention capacity of the sand mandrel prevents heat dissipation. The temperature rise locally decreases the flow stress of the material.

3.5 ECAP FOR POWDER CONSOLIDATION

At the industrial level, direct extrusion or compaction of even fine Al-based particles is limited due to extreme pressing pressure, and with no room for further increase of extrusion temperature due to structural coarsening and grain growth. Although utilisation of backward extrusion leads to a significant decrease in pressing load, on the other hand, it often yields insufficient straining and improper consolidation. Due to high pressing pressures (over 1.4 GPa) needed for breakthrough of the compacted material, direct extrusion was not possible below 350°C. During ECAP, the simple shear condition minimises the redundant normal forces, thus yielding a considerable reduction of pressing loads compared to conventional consolidation. In addition, it has been shown that the shear mode of plastic deformation leads to a change in the spherical pore geometry to an elliptical shape aligned with the shear plane, which is favourable for closure of voids under hydrostatic pressure. ECAP consolidation of UFG aluminium powders can be carried out at a lower (at least three times) pressing load compared to direct extrusion.

In order to consolidate the powder through ECAP, a can of suitable material is used to fill the powder. The encapsulated powder is ECAPed like conventional solid billets. The consolidation of powders by ECAP can be improved by applying back pressure during ECAP through the exit channel. An important advantage of imposing backpressure is that the hydrostatic pressure can be increased and is utilised for the compaction of powders. Technically, back pressure can be imposed in several different ways. The simplest way to increase back pressure within ECAP channels is to place a bulk front stopper in front of the compacted materials, as shown in Figure 3.8. The back pressure is induced due to the obstacle in the exit side, and this method is widely followed.

The drawback in this method is that the actual back pressure applied cannot be quantified directly. However, it can be calculated indirectly from the load required to deform the material in the channels with and

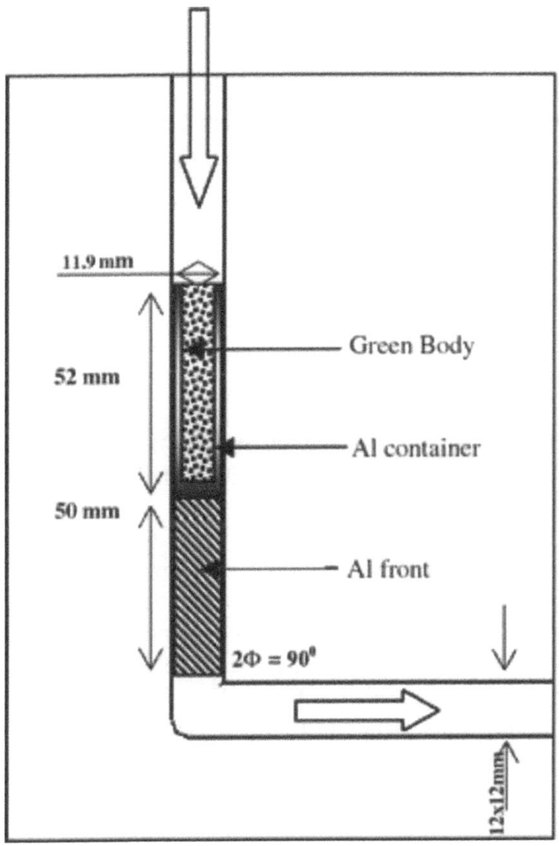

FIGURE 3.8 Schematic representation of imposing back pressure in ECAP using Al front stopper.

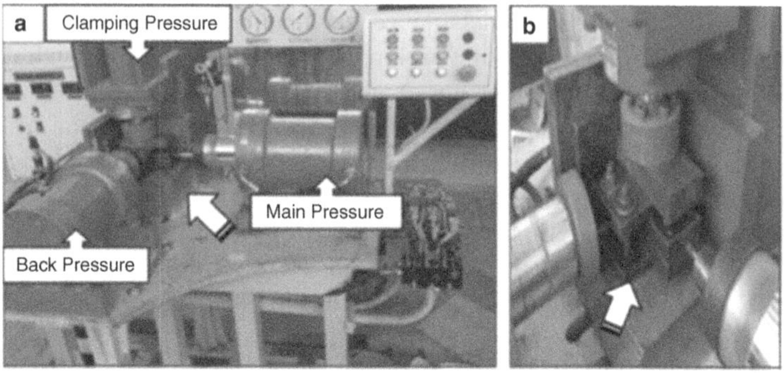

FIGURE 3.9 Photograph of (a) ECAP machine and (b) ECAP die set up with facility to apply back pressure.

TABLE 3.1
Relative Density of Al 5083 Powders Consolidated by ECAP without and with Back Pressure of 175 MPa

	Relative Density (%) After Three Passes	
Route	**Without Back Pressure**	**With Back Pressure of 175 MPa**
A	96	98
B_A	91	93
B_C	93	95
C	89	91

without a front stopper. The accurate way is to apply pressure from the exit channel. However, this needs a complicated die and experimental setup as shown in Figure 3.9.

The density values of mechanically alloyed Al 5083 powder consolidated by ECAP in various conditions are shown in Table 3.1. Among the four routes, the maximum density is obtained in the compacts processed in Route A. Since the strain is monotonous, plastic deformation leads to a change in the spherical pore to an elliptical shape aligned with the shear plane, which is favourable for closure of voids under hydrostatic pressure. Hence in Route A yields better density than in other routes. Almost the same trend is observed in most of the metallic systems.

The densification by ECAP is due to shearing exposed fresh and clean surface of the particles, and compressive stress ensures good contact between them. Under these conditions, bonding occurs spontaneously without needing high temperature and extremely high pressures, in contrast to conventional sintering, which requires a long time and temperature for bonding between particles to take place.

In addition to densification, ECAP refines the grains as well. The average grain size of conventional sintered is 26 microns, which is refined to 7 microns after two passes in Route A with back pressure. The ECAP processes increase both density and strength of the powder compacts as compared to conventional routes.

3.6 ECAP DEVELOPMENTS

ECAP has the biggest potential for scaling up since the development of microstructure after pressing does not depend on the initial size of the sample or the location within the sample. Processing larger sizes of bulk metals will certainly involve significant investment in tooling. Considering the above-mentioned facts, ECAP has experienced several alterations or modifications.

The major setback in ECAP is the friction between the billet and die walls. In order to reduce friction, movable walls in the channels were invented (Figure 3.10a). Die equipped with a cylindrical surface in the form of a rotatable roll was introduced to avoid friction (Figure 3.10b). An unequal channel extrusion in which the billet changes its cross-section in the outlet channel (Figure 3.10c) is used to avoid friction, but this modification of the outlet channel will increase the process force and decrease the tool life. Using the small slenderness of the billet or increasing the angle between consecutive channels in the die can help reduce the force. Most small design alterations have only a limited impact on improving the processing conditions. Thus, modifications that improve the performance of the ECAP method are of considerable importance to industry.

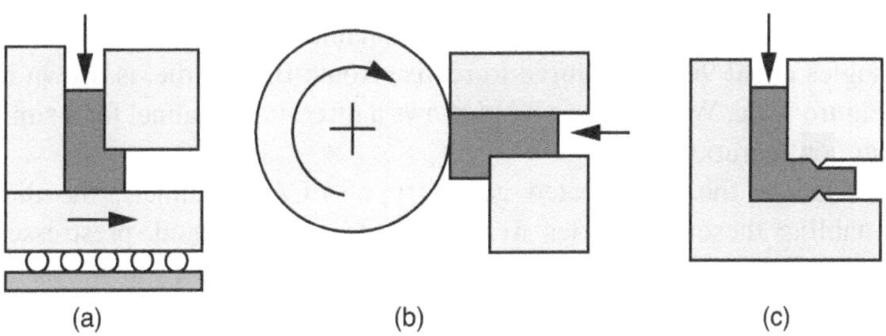

FIGURE 3.10 Modifications of ECAP: movable wall (a), rotatable roll (b), profiled output channel (c).

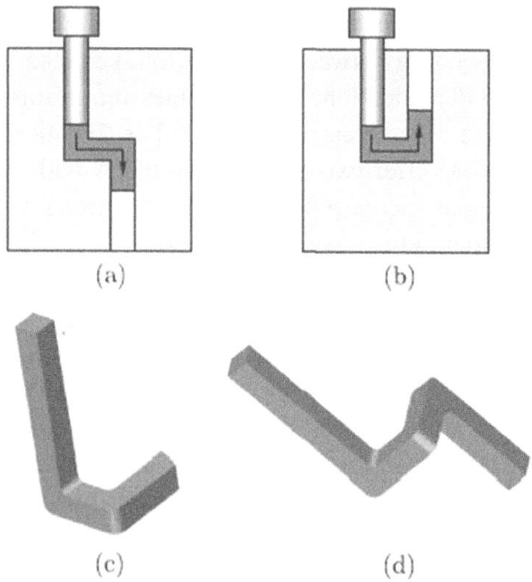

FIGURE 3.11 Channel designs for multi-turn ECAP: two-turn with Route C in die (a), two-turn with Route A in die (b), 2-turn with Route B_C in die (c) three-turn with Route B_C in die (d).

An idea to increase the billet's deformation during one pass in the die, giving a possibility of immediate input of the slug with reduced cross-section for the next pass in an S-shape channel, is shown in Figure 3.11a. This scheme is equivalent to the so-called Route C of ECAP, which involves rotating the billet about its axis by 180° between each pass in the classical "two channel" die. A two-turn, U-shaped channel, presented schematically in Figure 3.11b, is equivalent to Route A realised outside the die (no rotation of the billet). Using two-turn channels doubles the strain produced in one pass and increases the productivity of ECAP. In order to obtain homogeneous microstructures of equiaxed grains separated by high-angle boundaries, Route B_C (rotation by 90°) should be used. A square channel, with two turns and all angles equal 90°, configured to realise Route B_C in a die, is drawn in Figure 3.11c. While Figure 3.11d shows a three-turn channel for a similar configuration.

Due to the complicated geometry of ECAP channels, the dies enabling these geometries were designed as segmented, prestressed tools. Tool prestressing is the method routinely used in metal forming technology. Tool design and making is a well-developed sector that should facilitate industrial implementation of ECAP, even in the case of complex ECAP dies.

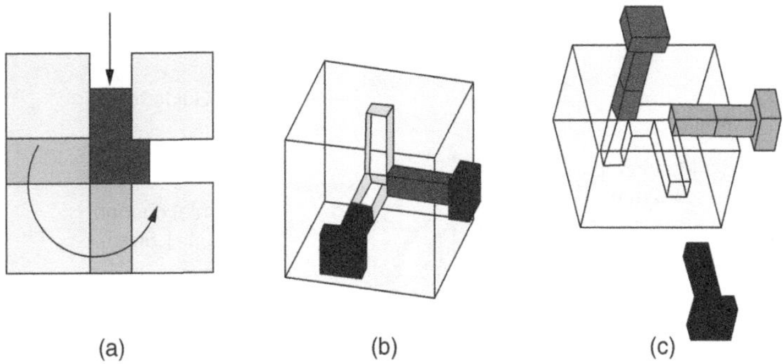

FIGURE 3.12 Multi-punch ECAP procedure: with rotating die (a), with multi-pass channels (b), with one-pass channels (c).

Manipulation of billets between consecutive passes of ECAP may cause problems in industrial situations. There is, however, a method of reducing the number of such manipulations. It can be achieved not only by increasing the number of channel turns in a die but also by increasing the number of inlets and outlets in a die and the number of corresponding punches. Figure 3.12a shows a cross-like configuration of channels with the billet material subjected to simple shear in the central hub at the channel crossing. The channels are presented to the appropriate punches by rotating the die before each deformation cycle. The deformation scheme realised is that of Route A with the leading end of the billet altered before each cycle. The billet is left in the die until the whole programme of deformation is finished. The ECAP die shown in Figure 3.12b features the same concept realised in a stationary die and the channels orientated in three dimensions. Thus, the deformation scheme is that of B_C with the leading end of the billet altered before each cycle. A similar die design for Route B_C without the leading end alteration is shown in Figure 3.12c.

In this design, the billet passes each channel only once, being successively pushed by the appropriate punches. The ECAP procedures presented in Figure 3.12 are in fact multistage metal forming processes realised in a single progressive die. The development and maintenance of shaping operations performed at one stand is very common for metal forming technology. It even seems reasonable to merge the fabrication of UFG metals with shaping products in the near future. The driving forces for the development of ECAP are the availability of machines (variety of presses), knowledge of tool design (e.g. prestressed dies) and good practice in billet treatment (lubrication techniques).

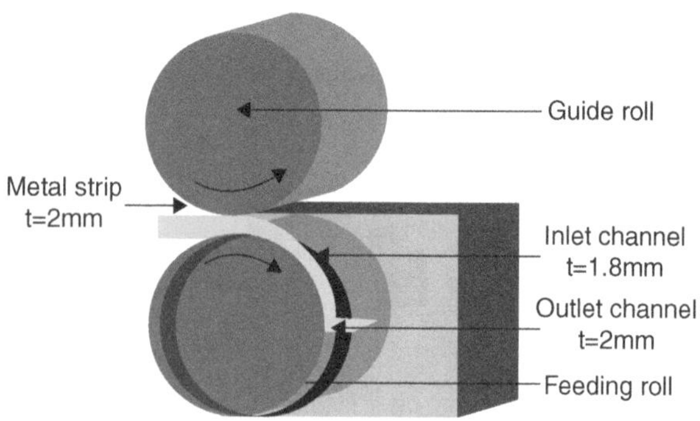

FIGURE 3.13 Schematic of ECAR device.

One of the major limitations in ECAP is the length of the billet. The length is restricted by the die dimensions. Equal channel angular rolling (ECAR), which is based on ECAP, provides repeated shear deformation on strip metals. The ECAR setup is illustrated in Figure 3.13. It was equipped with two rolls and two channels, whose thicknesses are dissimilar to each other. Firstly, the strip having an initial thickness of 2 mm is fed through the gap between the twin-roll and reduced into the 1.8-mm-thickness, according to the twin-roll gap, and the strip is entered into the inlet channel. The gap of the twin-roll is set as 1.8 mm in order to feed the strip into the channel by the friction force between the strip and the rolls. Then, the strip continuously passed through the forming zone, where the inlet and the outlet channels intersect, and exited through the outlet channel. The length of the product is not a constraint. The same setup can also be used for plates, rods and billets. This is also called Continuous ECAP.

3.7 ADVANTAGES AND LIMITATIONS

Equal channel angular pressing process holds the capacity of enhancing the mechanical properties like yield strength, ultimate tensile strength, hardness and ductility of majority of metals, alloys and composite materials through grain refinement and increased dislocation density. This method is useful in modifying the grain structure of the materials without altering its shape and size, which can help with the fabrication of complex components and reduce lead times for custom designs. The workpieces could be thin rods, flat plates, thin sheets, cylindrical parts or even workpieces with square sections. ECAP is a simple and cost-effective method of severe plastic deformation. ECAP, in particular, is one of these processes that also addresses the possibility of use on an industrial scale.

However, ECAP also has some limitations. The conventional ECAP process can cause the extruding ram to buckle, and so the design of the ram and punch is crucial. The conventional ECAP process can cause surface defects due to high friction. However, modifications of ECAP such as movable wall, rotatable roll and profiled output channel can reduce the friction and minimise surface defects. As compared to High pressure torsion (HPT), the amount of plastic deformation caused by the ECAP is less and grain refinement is less. The workpiece must experience multiple passes through the die in order to achieve a large amount of plastic deformation and every time the sample must be removed from the die in conventional ECAP. However, multi-passes in single die are possible in multi-turn ECAP dies, though designing such types of dies is critical.

4 High-Pressure Torsion

4.1 PRINCIPLES AND PROCESS OF HIGH-PRESSURE TORSION

The principle of high-pressure torsion (HPT) is shown in Figure 4.1 A thin disc is squeezed between two anvils (A and B) under a high pressure, P, and intense shear strain is introduced by rotating the two anvils with respect to each other. In the beginning, the material flows out laterally, and the thickness is reduced until the hydrostatic pressure is balanced with the friction. At this point, the material experiences real hydrostatic pressure. Normally, the lower anvil is fixed, and the shear strain is imposed by rotating the upper anvil with respect to the lower anvil by means of a motor and a large diameter gear. Maximum processing temperature of 1000°C under a pressure of 5 GPa is possible in HPT using the Tungsten carbide (WC) anvils and an induction heater coil. It is possible to apply pressure more

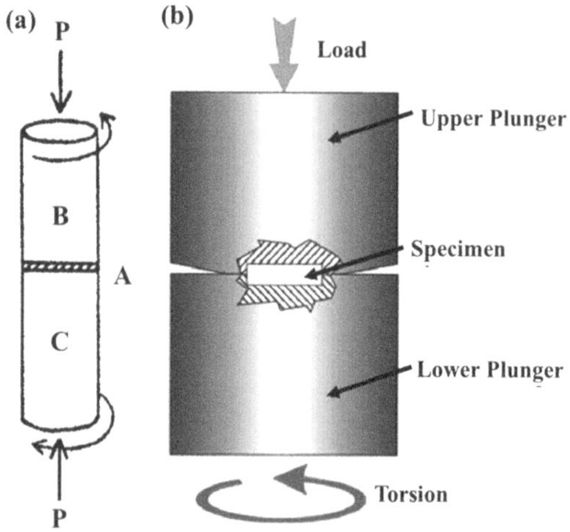

FIGURE 4.1 Principle of HPT in (a) ideal condition and (b) quasi-constrained condition.

DOI: 10.1201/9781003596110-4

than the yield strength of anvils if proper supports from surrounding parts are used.

It should be noted that the pressure distribution is not uniform in the disc sample during HPT processing because of the dimensional effects, friction effects and flow of material. It is suggested that the distribution of pressure becomes more uniform with an increased number of rotations. The pressure can be reasonably uniform during HPT if the disc sample is thin and the geometry and materials of anvils are selected carefully. Recent finite element simulation results showed that the pressure is reasonably isotropic in HPT, but it decreases with increasing distance from the disc centre. HPT using ring specimens with a simple form of anvils, having a ring-shaped groove on the surfaces, has been recently used for processing different metallic materials.

4.2 SHEAR STRAIN AND EQUIVALENT STRAIN

The shear strain, γ, during HPT is given by the following equation, provided that the thickness of disc remains constant during the deformation.

$$\gamma = r\theta/h$$

where r is the distance from the centre of disc, $\theta = 2\pi N$ is the rotation angle in radian (N is the number of turns) and h is the thickness of disc. The linear strain (e) is related to shear strain (γ) as $e = \gamma/\sqrt{3}$ as suggested by von-Mises and currently appears to be the most accepted physical mathematical equation among the SPD community. Application of a logarithmic equivalent strain ($\varepsilon = \ln(\gamma)$) for torsion rather than a von-Mises equivalent strain has been motivated by two main reasons. First, the geometrical changes at large strains are similar for shear and tension. Second, the difference between the stress–strain curves in tension and the stress–strain curves in torsion is very significant when the von-Mises equivalent strain is used.

4.3 EFFECT OF MATERIAL AND PROCESSING PARAMETERS

4.3.1 Effect of Melting Temperature on Grain Refinement during HPT

The grain size increases monotonically with increasing homologous temperature (i.e., with decreasing the melting temperature) as the recrystallisation temperature depends on the melting point directly. For metals with low melting temperatures such as Zn, Bi, Pb, Sn and In, the grain refinement is minor, while the grain size decreases significantly for metals with high melting temperatures. Interestingly, for Si and Ge, having atomic bonding features similar to minerals, the grain refinement by HPT processing is even more significant.

4.3.2 Significance of Temperature Rise and Anvils Alignment during HPT

Temperature may rise during HPT processing because of the severity of plastic deformation. Using standard heat transfer equations, it is found that the maximum temperature rise with a rotation speed of 2.6 rpm may reach 37°C within 5 seconds. However, the real temperature rise must be much smaller, because actually heat flows from the sample three-dimensionally into massive anvils and plates. So, the thermal factor thus appears to be of absolutely no importance. Finite element simulations confirmed that the temperature rise during HPT processing is not significant when compared to the melting temperature of metals.

4.3.3 Thickness of Sample and Strain Distribution

The thickness of disc during HPT processing significantly influences the distribution of pressure and strain. When the disc is thin, all parts of disc are under hydrostatic pressure except for a narrow area around the edge. The ratio between thickness and diameter should be less than 0.1 to have a uniform distribution of strain and pressure.

4.3.4 Slippage during HPT at Low Pressures

Since the invention of HPT, an important concern is the significance of any frictional slippage that occurs between the sample and the anvils during HPT processing. Many observations at low pressures are due to the effect of slippage. Two different regions are observed in the plots of steady-state shear stress against pressure, as shown in Figure 4.2.

In the beginning, the shear stress increases rapidly with increasing pressure where the internal shear stress is below the yield stress and only the surface slippage occurs. Subsequently, the shear stress increases slowly with increasing pressure where the internal shear stress exceeds the yield stress, internal plastic flow occurs (material sticks to the anvils without any surface slippage). There is, however, a gradual transition from the surface slippage to internal plastic flow,

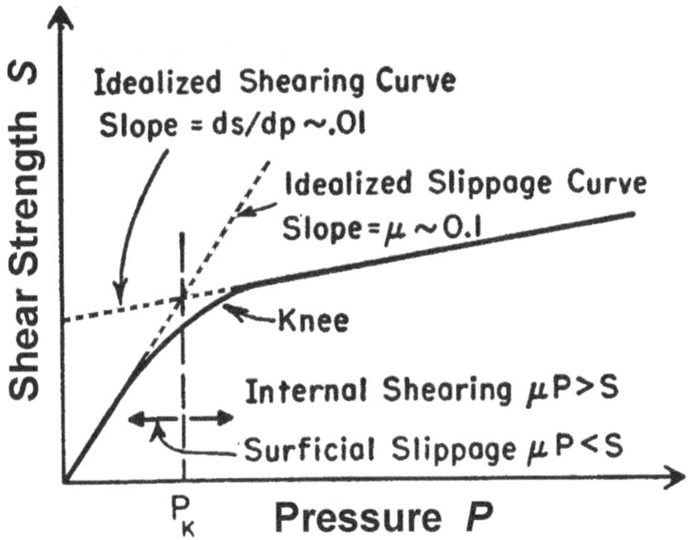

FIGURE 4.2 Schematic relation between shear strength and pressure.

and this transition region is called knee. Many researchers suggest that the pressure should be higher than 3.5–5 GPa to avoid any slippage. Figure 4.2 suggests that the minimum pressure to avoid slippage should be slightly higher than the maximum steady-state region (knee).

The effect of slippage is usually considered negligible on HPT because of the roughness of the surface of the anvils. Direct measurements of slippage during HPT processing showed that the extent of slippage is reduced by increasing the pressure, and it disappears at pressures higher than a critical value. The value of critical pressure is higher for the harder materials and somewhat higher at higher rotation speeds. It is also important that the friction coefficient for the upper anvil and the lower anvil should be the same to avoid heterogeneity in the shear strain.

4.3.5 Significance of Hydrostatic Pressure

It is a known fact that hydrostatic stresses decrease normal components of stress without affecting shear stresses resulting in improved ductility of the materials. The physical reason for this increase in ductility is that the external hydrostatic pressure does not permit the atoms to separate easily, and thus, the fracture and propagation of cracks are suppressed significantly. It should be noted that hydrostatic pressure should be compressive to get enhanced ductility in brittle materials. For example, under ordinary conditions, it is necessary to heat sapphire to 2000°C before it can be deformed plastically, while under a hydrostatic pressure of 2 GPa, sapphire tolerated very large plastic deformation not only in shear but also in simple compression. Cast iron, which is brittle under ambient pressure, became ductile under high hydrostatic pressures. Many metallic materials exhibit well-enhanced plasticity at room temperature under high hydrostatic pressures and many substances, even brittle ones, could be processed by HPT to large shear strain without failure when HPT is carried out in hydrostatic conditions.

An increase in torque with hydrostatic pressure is reported in recent studies on HPT processing in the quasi-constrained condition. Although few studies reported that the final steady-state microstructure (especially the concentration of vacancies) and mechanical properties depend on the hydrostatic pressure, many other studies reported that the steady-state microstructure and mechanical properties are reasonably independent of applied hydrostatic pressure when the pressure is higher than a critical value.

4.3.6 Significance of Homologous Temperature and Rotation Speed

The effect of rotation speed (strain rate) on the shear strength of materials processed by HPT strongly depends on their "melting points and capacity for recrystallisation", as shown in Figure 4.3

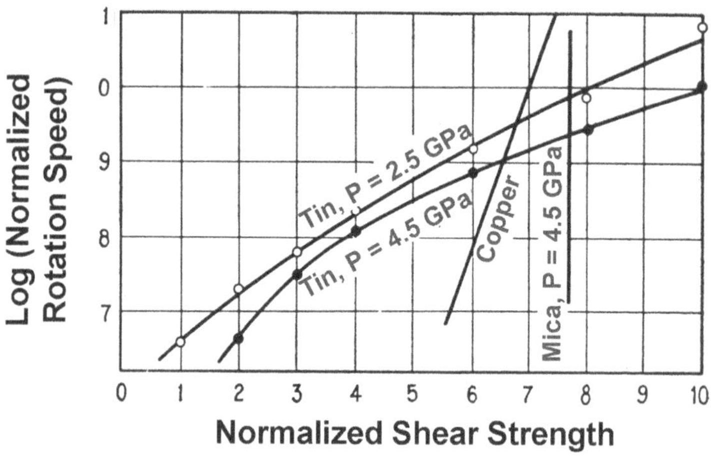

FIGURE 4.3 Variations of steady-state shear strength as a function of rotation speed for several materials during processing by HPT.

which reveals that for Sn and Pb with low melting temperatures, the shear strength significantly increases with increasing rotation speed, whereas the shear strength of Ag and Cu with moderate melting temperatures is less sensitive to the rotation speed and the shear strength of mica mineral with high melting temperature is independent of the rotation speed. It is also suggested that not only the shearing speed but also the pressing speed should be controlled during the HPT processing; otherwise, it can influence appreciably the properties of some materials, such as glasses.

The results indicate that the homologous temperature is one of the most important parameters in the HPT processing of metals. The drastic changes in the microstructure and flow stress at homologous temperatures higher than 0.3 is due to the significant effect of recrystallisation. It should be noted that for alloys, dislocation-solute atom interactions play a more important role than the melting temperature in the evolution of microstructure and plastic flow.

4.4 HPT FOR POWDER CONSOLIDATION AND SYNTHESIS OF ALLOYS

HPT has a great potential to synthesise densely packed materials from mixed powders without a sintering process (Figure 4.4). It is also possible to synthesis supersaturated solid solution. For example, it is possible to mix more Fe in the Al matrix using HPT in the cast Al–Fe alloy; there are possibilities to prepare supersaturated Al–Mg alloys by mixing Al and Mg powders. The addition of Mg to Al is popular in the Al industry because Mg is a typical element to introduce significant hardening of Al through a solid solution effect. Although the equilibrium solid solubility of Mg in Al is about 1 mass % at room

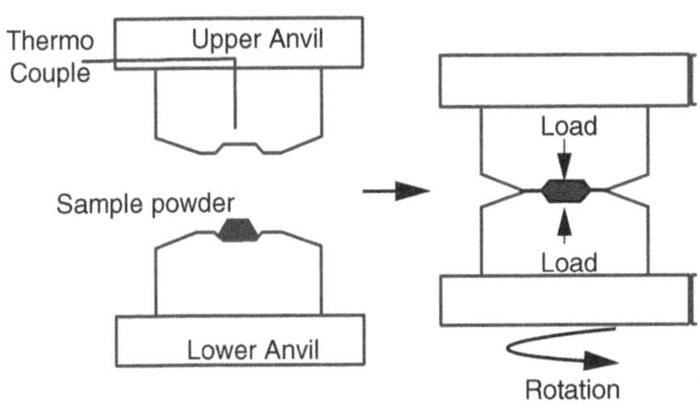

FIGURE 4.4 A schematic illustration of HPT used for powder consolidation.

temperature, an extended solid solubility of even up to 30 mass % of Mg in Al can be achieved by rapid solidification as well as by HPT of elemental powders.

In HPT, high-purity micron-sized powders are mixed with various amounts. The required amount of powder mixture is placed in a shallow circular hole, located at the centre of the lower anvil of the HPT facility. The lower anvil is lifted until it contacts the upper anvil, which has the same dimensions as the lower anvil, and then rotated with respect to the upper anvil. This HPT operation will be at room temperature with an applied pressure of 2.5–6.0 GPa for different numbers of revolutions. The rotation was initiated for 5 seconds after the load application, and the temperature rise during the HPT operation is monitored using a thermocouple placed 10 mm above the sample. The thermocouple is guided through a hole made vertically in the centre axis of the upper anvil. The measurement indicated that the temperature gradually increased up to 353 K after 100 turns. The final sample after the HPT operation is in the shape of a disc with dimensions of a circular hole. It should be noted that an extra amount of the mixed powder will be deliberately added to prevent the upper and lower anvils from contacting each other directly so that the damage and wear to both anvils will be avoided. The XRD patterns of the alloys clearly indicate that the alloying can be achieved from the mixed powders and the supersaturation as well as the new phase formation occur by the process of HPT for alloy compositions much higher than the equilibrium composition.

4.5 HPT DEVELOPMENTS

Because of the small size of the sample in the HPT processing, the method is considered inappropriate for industrial applications when compared to the other SPD methods. There have been a few attempts to process bulk samples with a rod shape by modifying the HPT anvils. Attempts made to process sheets/rods of metallic materials in a continuous manner are briefly discussed in the following sections.

4.5.1 Continuous High-Pressure Torsion

In the continuous high-pressure torsion (CHPT), the sample is processed in an expeditious way so that the sample size is increased without sacrificing the pressure. The facility for the CHPT, as schematically illustrated in Figure 4.5, consists of two anvils. The lower anvil, which is rotated during the process, has a flat surface with a roughened

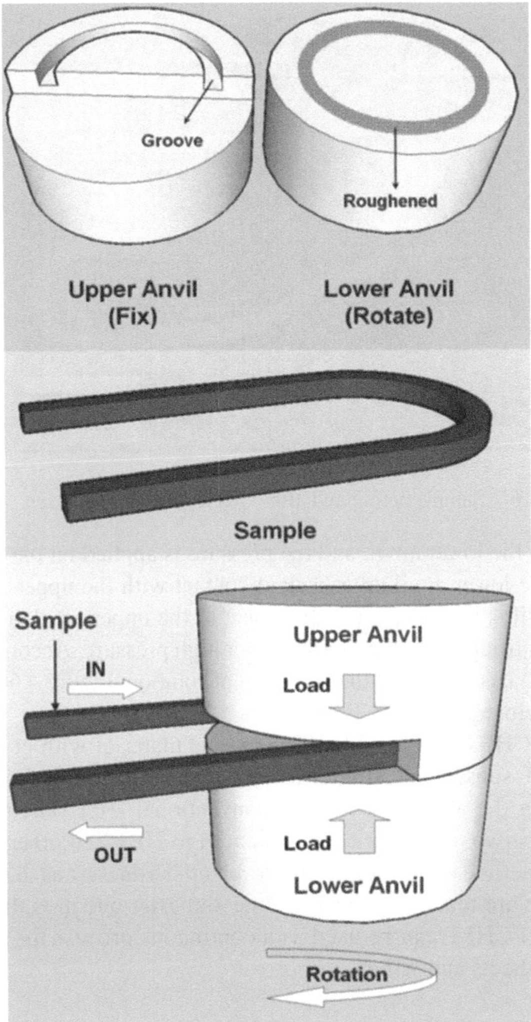

FIGURE 4.5 Schematic illustration of CHPT.

ring-shaped area, while the upper anvil, which is fixed during the process, has a half-ring-shaped groove on the surface.

The surface roughness of the upper anvil is less with respect to the surface roughness of the lower anvil so as to induce a continuous flow of the material due to the difference in slippage. A U-shaped specimen, as shown in Figure 4.5, can be used as an initial sample. Each sample is

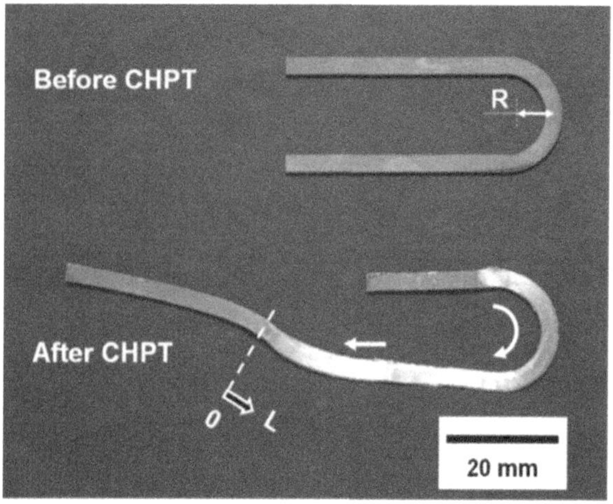

FIGURE 4.6 Sample before and after CHPT for one revolution.

placed on the lower anvil, and the pressure is applied on the sample by raising the lower anvil up to a rigid contact with the upper anvil. The lower anvil is then rotated with respect to the upper anvil, and a shear strain is introduced in the sample under high pressure. Accordingly, the material starts to flow in the direction of rotation. Figure 4.6 shows the sample before and after CHPT.

Thus, CHPT can provide a continuous material with grain refinement and subsequent strengthening in metallic materials without sacrificing the advantages of the conventional HPT. The method of CHPT has two main merits as compared to HPT and other SPD processing methods: (i) a saturated level of hardness and a minimum grain size are reached in a short time and after one pass through the anvils; (ii) CHPT can be used as a continuous process for producing metallic sheets and wires.

4.5.2 Incremental High-Pressure Torsion

In incremental high-pressure torsion (IHPT), similar to HPT, material is deformed between two anvils, which are loaded with pressures in the range of several gigapascals. Here, however, the anvils have a drilling hole through them. The inner hole of the anvils is slightly conical (Figure 4.7a), so that the diameter of the hole is somewhat larger at the top of each anvil than on the bottom. The cylindrical specimen is inserted into the anvils and the local position of the specimen within the anvils is confined by the height of two additional cylinders, called support cylinders, inserted from the outside.

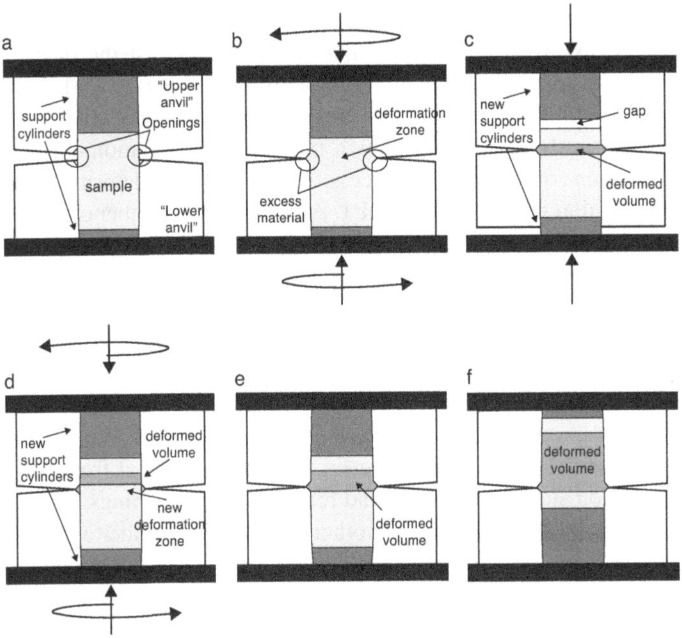

FIGURE 4.7 Schematic of incremental high-pressure torsion (IHPT) illustrating some significant processing steps. (a)The unloaded setup. (b) Anvils with loaded specimens ready for shear deformation. (c) Shift of deformed material through the change of the support cylinders. (d) Next deformation step. (e) Deformed volume after the second deformation step. (f) Final stage of the deformed specimen after several deformation steps.

In the unloaded state, the slight conical shape of the upper drilling hole prevents the upper anvil from contacting the lower anvil. This is because the specimen can only move into the upper anvil as far as the diameter of the specimen is smaller than the inner diameter of the anvil, which becomes smaller further as the specimens move into the anvil. At the top of each anvil, there is also a small opening to concentrate the shear deformation in the deformation zone, and this opening also ensures that excess material can flow out, as in the case of classical HPT anvils. When the specimen is loaded, the entire specimen also becomes slightly conical according to the shape of the anvils and the free-standing material between the anvils bulges, gets compressed, and the excess material flows out of the openings as illustrated in Figure 4.7b. A further displacement of the specimen inside the anvils is suppressed by the support cylinders. The excess material confined between the anvils generates the hydrostatic pressure component due to the evolving friction and also prevents the anvils from touching each other and causing excessive wear. Then, one anvil is rotated against the other. The presence of the compressive force, combined with the conical shape of the anvil and specimen, induces tangential friction forces that grip the specimen in the anvils and allow a free shear deformation only in the deformation zone during the process. At this stage, the deformation state is very similar to conventional HPT. After a certain number of rotations, the specimen is unloaded. Then a different pair of support cylinders is inserted, where the lower one has a larger height than the top support cylinder. Because of this, there is a small gap between the support cylinder and the specimen (Figure 4.7c). When the pressure is applied, the lower support cylinder pushes the specimen upwards to fill the gap. At the same moment, the upper and the lower anvils move together downwards, and the new undeformed material moves into the deformation zone (Figure 4.7d). However, the excess material from the first deformation step cannot move and remains in the openings. Then, one anvil is again rotated against the other for a certain number of rotations, Figure 4.7d, and the new undeformed material located in the deformation zone gets deformed. Due to this incremental process, the deformed volume increases, Figure 4.7e. By doing this process multiple times and by adequately changing the dimensions of the support cylinders, the volume of the deformed material increases continuously, Figure 4.7f. The growth of the deformed material zone is controlled by the height change of the support cylinders between consecutive deformation steps and defines a certain step size which is smaller than the thickness of the deformation zone evolving in one deformation step to ensure a distinctive overlap of the deformation zones between the following deformation steps. The support cylinders can also be replaced with discs, which can be stacked together to a certain height so as to avoid the necessity of a large number of different, individually manufactured support cylinders. Thus, IHPT combines the advantages of conventional HPT and overcomes the limitation of a restricted specimen thickness.

4.5.3 High-Pressure Torsion Extrusion

High-pressure torsion extrusion (HPTE) is based on a combination of the conventional high-pressure torsion technique and extrusion. During HPTE, a specimen is extruded through sectional containers rotating relative to each other. The specimen is subjected to shear deformation in the area where the containers meet each other. In HPTE, material is extruded through two containers (Figure 4.8), with one of them being stationary and the other rotating normally.

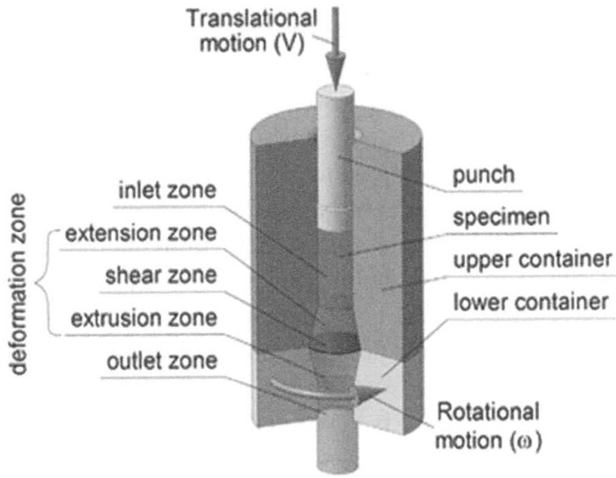

FIGURE 4.8 Schematic diagram of the HPTE process.

Initially, the channels of all containers have to be filled completely with material. This can be achieved by either using a specially shaped initial specimen or by closing the outlet with a plug. Once both channels are filled, the plug is removed. The extruded material then acts as a plug and ensures that the two containers remain fully filled during the process. The reduction of the channel diameter and friction results in forces acting opposite to the extrusion direction and creating hydrostatic pressure in the shear zone. Therefore, a specimen is deformed by the punch (moving with the velocity v) as well as by the containers (one of them rotates with the velocity ω). Consequently, the entire length of a specimen is deformed in torsion as the specimen gradually passes through the shear zone. Therefore, the resulting microstructures and properties are homogeneous in the longitudinal direction. The geometry of the containers can be modified to achieve the required level of hydrostatic pressure in the shear zone.

After a single extrusion pass, a high accumulated strain is induced in the specimen, which is the main advantage of the HPTE process. Furthermore, the presence of a strong velocity gradient in the specimen cross-section during HPTE provides the possibility to process hybrid materials or composite parts with functional elements.

4.6 ADVANTAGES AND LIMITATIONS

Among the various SPD processes, HPT is effective for introducing extremely large shear strains for grain refinement. As an important feature, it involves considerably high amounts of hydrostatic pressure, which inhibits crack formation during deformation and leads to almost no limitation in the applicable strain. Consequently, this technique also allows HPT of intermediate and high-strength starting materials, such as steels and tungsten, even at low homologous temperatures. Furthermore, HPT is also fairly simple, making it a quite reliable and cost-effective method compared to other SPD processes.

However, the HPT process has three main limitations: first, the sample shape is in a form of disc which is not appropriate for many industrial applications compared to wire or sheet forms; second, the sample size is generally limited to a maximum of 35 mm in diameter in order to keep a high applied pressure; third, as the strain is generated in proportion to the distance from the disc centre, an inhomogeneous distribution of microstructure is developed across the diameter. The first two limitations can be overcome to a certain extent by using advanced HPT processes such as CHPT, IHPT and HPTE where relatively large samples can be used. However, continuous production irrespective of length is yet to be realised. To solve this, efforts are being carried out to combine HPT and conform processes.

In order to overcome the third limitation, the HPT using a ring shape, which is more appropriate than a disc shape for industrial applications, is used with a simple modification of the conventional anvils. Using the ring HPT, it is possible not only to eliminate a less strained and coarse-grained central part but also to scale up the sample up to 100 mm in diameter.

5 Twist Extrusion

5.1 PRINCIPLE AND PROCESS

The key idea in twist extrusion (TE) is based on pressing out a pseudo-prismatic specimen through a die with a longitudinal profile consisting of two prismatic regions separated by a twisted part. Each cross section of a billet undergoes severe shear deformations at the distorted region, such that first, it is twisted by being deformed at a given angle in one certain direction, and then is re-twisted at the same angle in the opposite direction (Figure 5.1).

TE is also performed under high hydrostatic pressure, which is created by applying backpressure to the specimen in the exit channel, similar to equal channel angular pressing (ECAP). It is possible to produce more isotropic and homogeneous deformation by turning the samples 90° in each consecutive deformation or alternatively, make the use of consecutive clockwise – anticlockwise – clockwise twists. This is very important for electronic and magnetic materials. However, billet rotations between TE passes have no significant effect on the plastic flow, like in ECAP passes. This is due to the axial symmetry of the process. The samples used in TE are normally wrapped with Teflon tape and also silicon sprayed to reduce friction.

In TE, twisting of the specimen occurs due to the special form of the extrusion die (Figure 5.2). Since the main dimensions of the work piece remain unchanged when it exits the die, it is possible to repeat the operation as many times as required to achieve the higher levels

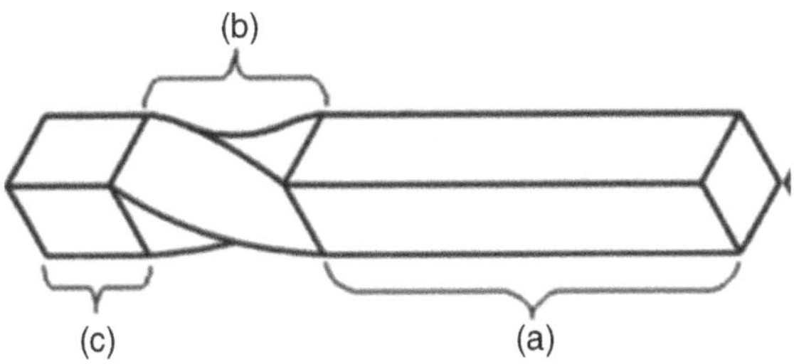

FIGURE 5.1 Input and output sections and guides for twist extrusion die: (a) input guide, (b) twisted channel and (c) output guide.

DOI: 10.1201/9781003596110-5

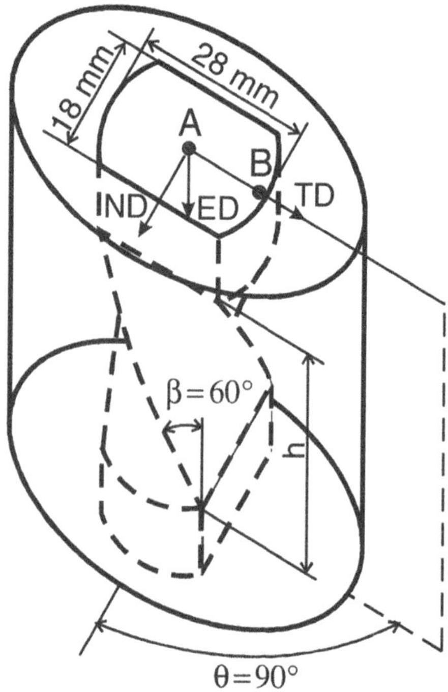

FIGURE 5.2 Twist extrusion die.

of plastic deformation and providing the conditions of controlling the microstructure and properties of the material.

5.2 SHEAR STRAIN AND EQUIVALENT STRAIN

In TE, unlike axial loads as in the case of ECAP and HPT, which produce a uniform, or average, strain over the cross section of the object, TE creates a distribution of stress over the cross section. When a twist is applied to the structure, it will twist along the long axis of the rod, and its cross section remains the same. For the solid cross-section shaft with material homogeneity on the cross section, both the shear strain and shear stress vary linearly with radial position on the cross-section. The shear strain, γ (gamma) is $=(r\times\beta)/h$

Where r – radius (distance from the centre), β = angle of twist in radians and h is the length (height) of the sample subjected to twist (Figure 5.2).

The shear strain is subsequently expressed in dependence on the particular cross-sectional location within the processed material,

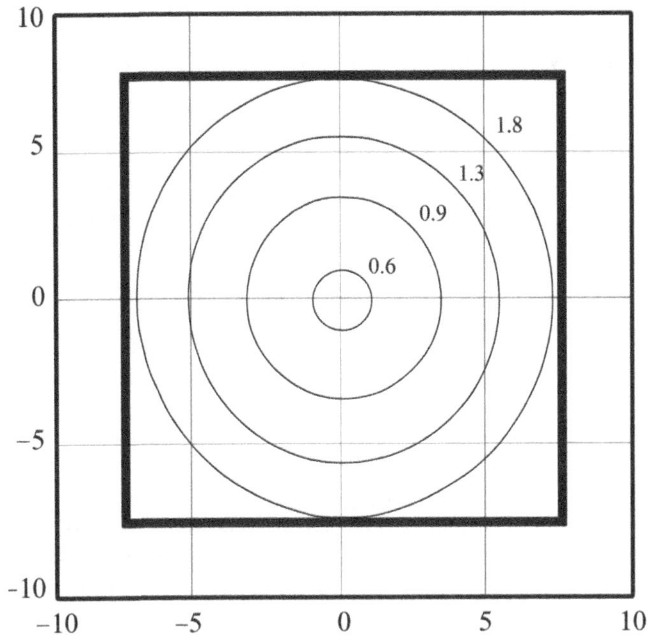

FIGURE 5.3 Distribution of strain after one pass of TE.

regarding axis – cross section periphery of the sample- i.e., strain value in radial direction increases from the centre of the cross section towards the edge (Figure 5.3).

Thus, the microstructure is more homogenised in the lateral regions of the cross section than in the central regions.

5.3 EFFECT OF PROCESSING PARAMETERS

5.3.1 Effect of Geometry of Die

Strain distribution at deformation zones is strongly dependent on the geometry of the die profile, that is, deviation angle (θ) and rotation angle (β), and one can change the strain intensity at different deformation zones by changing these factors. Therefore, the mean grain size in the lateral regions is reduced by about 60% after one pass, while in the central region it is reduced only by 26%. Also, the grains in the lateral regions will be mostly elongated, while they look coarsened in the central region. Also, by increasing the number of passes, the grain size

FIGURE 5.4 Grain size heterogeneity index for different passes.

differences in the various parts of the cross section are decreased, and the microstructure becomes more homogeneous (Figure 5.4). This is due to the stabilisation of microstructure and the durability of properties once strain exceeds its highest level. This is typical characteristics of all deformations based on simple shear, such as ECAP.

Therefore, despite heterogeneity of deformation, the structure and properties of the material tend to be homogeneous in the higher passes of TE.

5.3.2 Effect of Processing Main Routes of TE

There are two types of twist dies: clockwise (CD) and counterclockwise (CCD). When transitioning from CD to CCD, the shears in each of the four deformation zones reverse its sign. This gives us two main routes of TE:

Route I: CD+CD (or CCD+CCD),
Route II: CD+CCD (or CCD+CD).

The Route I is similar to Route B_C and Route II is similar to Route B_A of ECAP. Figure 5.5 schematically shows the sign-sensitive change in the shift deformation along the transversal (Plane T) and the longitudinal (Plane L) planes during TE along Routes I and II. In the transversal plane, Route I gives a cyclic deformation with some amplitude A, while Route II gives a cyclic deformation with amplitude 2A. In the longitudinal plane, Route I gives a monotonic deformation while Route II

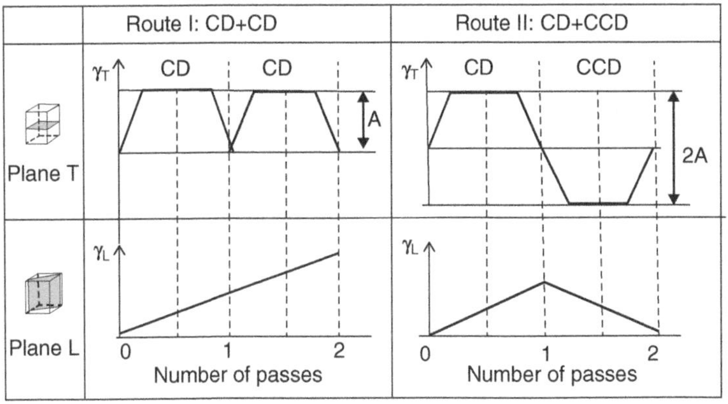

FIGURE 5.5 Sign-sensitive change in the shift deformation along the transversal plane (T) and the longitudinal plane (L) during twist extrusion along Routes I and II.

gives a cyclic deformation. Different loading paths can lead to different structures and properties.

5.4 DEVELOPMENTS IN TE

The major limitation in conventional TE is the length-to-diameter ratio. It should avoid the buckling stresses and the length of the punch, which needs to be protected against buckling. As in TE, the surface stresses are high, the appearance of surface cracks in relatively ductile materials could not be avoided, and, of course, friction is another limiting factor. The following processes eliminate any one of the limitations or all the limitations.

5.4.1 Twist Channel Angular Pressing

Twist channel angular pressing (TCAP) was first introduced by Kocich et al. in 2010. TCAP is inspired by two individual severe plastic deformation (SPD) techniques, ECAP and TE, combining the positive aspects of both into a unique novel method. ECAP and TE can be combined in several ways, as shown in Figure 5.6a–c. Nevertheless, as experimentally proven, only one of the variants introduces reasonable processing parameters (e.g., punch load, temperature distribution) and preserves the constant shape of the sample, thus ensuring the repeatability of the entire process; Variant I (Figure 5.6a).

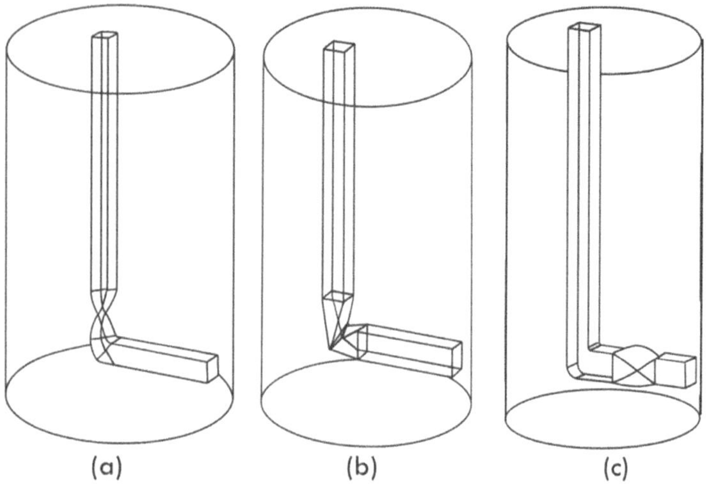

FIGURE 5.6 Experimentally investigated variants of twist channel angular pressing (TCAP): (a) Variant I, (b) Variant II, and (c) Variant III.

Although Variant II (Figure 5.6b) also combines both the ECAP and TE processes, this variant brings about several issues, probably the most severe of which is that positioning the twist directly to the bending introduces a substantial local reduction of the channel cross section. In other words, the process then locally involves forward extrusion, which necessarily changes the processed sample cross section and introduces changes in its dimensions, by which the repeatability of the process, the crucial factor of all SPD processes, is disabled. This phenomenon can only be compensated for by the application of backpressure. The second significant issue for this variant is a substantial increase in the punch load, which becomes even more substantial when applying backpressure. Among others, numerous cross-sectional reductions typical for this variant introduce a complex stress state, which can be unfavourable for the processed material.

The number of issues is reduced for Variant III (Figure 5.6c), although they are not eliminated. The main issue is related to the location of the twist being positioned behind the bending, which leads to a partially twisted final shape of the processed sample. Besides, by locating the twist behind the bending, the issues of the TE method, that is, partial cross-sectional deformation of the processed sample, as well as cross-sectional inhomogeneity of the imposed strain, are not eliminated; unlike Variant I, Variant III does not feature any final deformation zone that could suppress/reduce the strain inhomogeneity imparted by the twist.

The increase in temperature for Variant II is more than twice as high as for Variant I, which affects the structures of processed materials featuring low melting temperatures. Moreover, the increase in temperature introduced by the development of deformation heat does not correspond to the increase in the imposed strain, which is only approximately 20%. For these reasons, only Variant I is more preferable than the other variants.

The total strain in one pass of TCAP is equivalent to the geometrical addition of two individual processes – ECAP and TE. Thus, the total strain imposed is very high and results in fine grains and better mechanical properties with less number of passes and less processing time. In addition, the advantageous combination leads to homogenisation of final properties throughout the sample cross section and length, while keeping the shape unchanged.

5.4.2 Planar Twist Extrusion

The sample size for TE is limited by two factors: the aspect ratio (length/diameter) has to be smaller than a critical value to avoid bending of the press plunger during extrusion, and the plunger has a limited travel distance. Traditional extrusion and ECAP possess similar limitations, which are overcome by the application of two-sided gripping of feedstock as well as containers and dies with movable walls. The latter permit control of the magnitude and direction of the frictional forces between the sample and the tool, relief applied to the plunger pressure by partial load transfer onto the movable die walls, designing the plunger as a more stable and stronger component. In the die design in which two lateral sides of the sample are surrounded by the plunger and the bottom side of the channel is free to move in order to reduce the frictional forces. The other suggested design is using a plunger surrounding three lateral sides of the sample but not having the bottom gliding part. The latter design is more advantageous as it not only allows for reducing the frictional forces, but also prevents the plunger from bending.

The same design concepts cannot be applied to TE, as the die channel is bounded by twist, i.e., non-flat walls. TE with two parallel die walls, named Planar Twist Extrusion (PTE), has, therefore, been proposed to overcome these shortcomings. The two processes differ in the shape of the die channel: TE has a screw shape, while the two opposite walls of the PTE dies are flat.

The die channel for PTE had three distinct segments, namely entry, deforming and exit. The entry and exit segments had a rectangular cross section, while the middle part of the die, i.e., the deforming segment – cross section varying along the die axis from rectangular to parallelogram and then back to rectangular. Schematics of the PTE set up is shown in Figure 5.7.

During PTE, the sample is pushed by the plunger through the die channel formed by two curved and two flat walls, the latter being made as a single part together with the plunger. At the place of loading, the plunger has H - beam cross section with an area larger than cross section of the sample being extruded. This feature allows to improve the stability of the plunger and increase the sample length. The load is transferred from the main and backpressure plungers to the sample itself through two channel-filling samples (CFS). The backpressure is applied to the front side of the sample so that it can fill the die channel. Moreover, this pressure also raises the hydrostatic pressure in the deformation zone, thereby improving metal plasticity and promoting grain refinement.

Figure 5.8 illustrates the stages of PTE. At the beginning of the process, the lower CFS is brought into the twist segment of the die (Figure 5.8a and b). Then the sample itself and upper CFS are inserted into the die (Figure 5.8c), and both of them are pushed through the twist channel. Following the extraction of the lower CFS and the sample

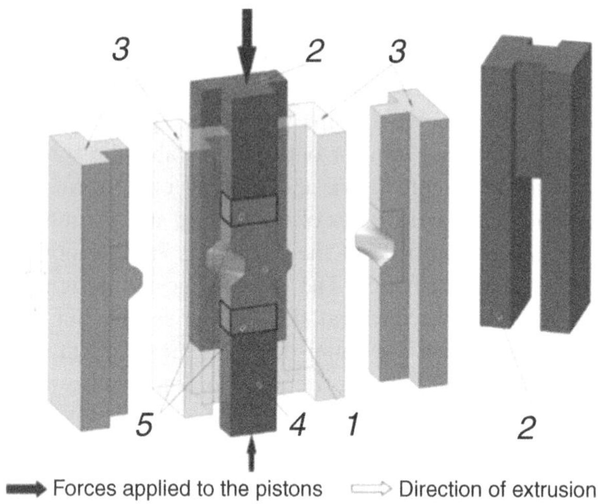

FIGURE 5.7 PTE die with two flat movable walls (1 – sample, 2 – main plunger, 3 – curved stationary walls, 4 – backpressure plunger and 5 – channel-filling sample).

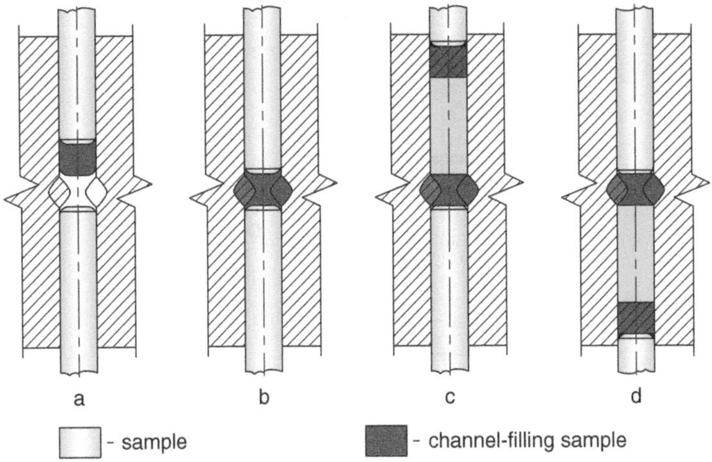

FIGURE 5.8 Stages of PTE process.

from the die, the remaining upper CFS now becomes the lower one for the next cycle (Figure 5.8d), and extrusion can be repeated.

5.4.3 Hydrostatic Twist Extrusion

The major problem in TE or in TCAP is the sample length and friction. PTE consists of a movable die and fixed punches. Although the sample length can be increased, the problem is not solved completely. Due to the short length of the processed sample, which is caused by high friction in the entrance of channels, the main challenge in TE processes is industrialisation. Increasing the sample length causes a higher probability of yielding and bulging of pressing punch. Using hydrostatic fluid pressure is one of the solutions to overcome the friction and the length limitation problem. Hydrostatic pressure leads to enhancement of sample workability, an increase in maximum applied plastic strain before fracture, and a significant reduction in total load.

A schematic of the Hydrostatic Twist Extrusion (HTE) method is depicted in Figure 5.9. The die setup consists of a die, backward punch, punch and seal. First, the sample is put into the die. Then, the fluid fills the gap between the sample and die, and its pressure increases by pressing the punch in the seal as shown in Figure 5.9. Using backpressure constantly is not required in this method, but it is better to use a backward punch outside the forming area to fill the corner gaps between the sample and the die. To start the process, the initial sample is pressed

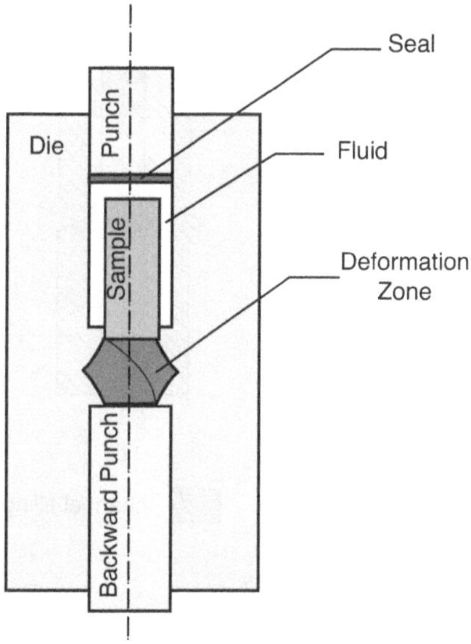

FIGURE 5.9 Schematic of HTE process.

into the deformation zone by the pressurised fluid caused by the punch movement, and the sample is twisted.

After the contact is made between the backward punch and the sample, the backward punch is removed, and the process continues. In HTE, load is independent of the length of the sample. The existence of fluid eliminates friction between the sample and die and reduces load significantly, which means that longer samples can be processed. The HTE process is suitable for the production of industrial-scale UFG samples.

5.5 ADVANTAGES AND LIMITATIONS

The unique advantage in TE is that the elongation to failure remained almost unchanged while the strength of the material increased. This is due to the inhomogeneous distribution of grain size, which is, in turn, caused by the formation of dislocation walls and subgrains in grains containing high dislocation density, converting them into smaller grains, while the grains with low dislocation density remain nearly unchanged. Thus, a considerable increase in strength does not decrease the formability of the material in TE. TE provides some advantages over ECAP in the extrusion of hollow parts and rectangular cross-sections.

The sample size for TE is limited by two factors: the aspect ratio (length/diameter) has to be smaller than a critical value to avoid bending of the press plunger during extrusion, and the plunger has a limited travel distance. This could be avoided in hydrostatic twist extrusion (HTE) by introducing suitable hydrostatic pressure to avoid buckling.

The fine-grained structures of the samples processed by TE, in some cases, are strongly anisotropic. Anisotropy, however, can be obviated by changing the die profiles, increasing the operating temperature and subsequent recrystallisation heat treatment or by applying deformations like drawing or direct extrusion after processing. Also, combining TE with ECAP, called Twist Channel Angular Pressing (TCAP) results in isotropic fine-grained materials without affecting the dimensions of the billet.

6 Cyclic Extrusion and Compression

6.1 PRINCIPLE AND PROCESS

During cyclic extrusion and compression (CEC), large deformation is imposed gradually on the processed materials by increasing the pass number. The amount of hydrostatic pressure increases with increasing plastic deformation and the level of sample hardening. This significantly facilitates material deformation conditions during CEC processes, especially in the hcp materials such as magnesium. Schematic representation of this process as well as its key die parameters are shown in Figure 6.1.

CEC die consists of two punches with a cylindrical cavity. The cylindrical-shaped die, which has two equal channels of diameter D, is connected in the middle by a smaller channel of diameter d. Each CEC pass applies an extrusion followed by compression on the billet in the main deformation zone (MDZ) so that the billet eventually returns to its original shape in the outlet channel. In this regard, the processing procedure consists of the following two steps. Initially, the billet

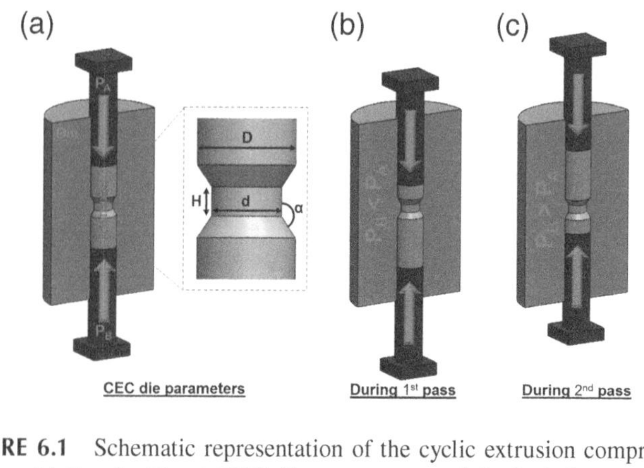

FIGURE 6.1 Schematic representation of the cyclic extrusion compression process: (a) the significant CEC die parameters and (b & c) the processing conditions of the first and second passes of the CEC process.

DOI: 10.1201/9781003596110-6

is placed on the upper channel and pushed by punch A to enter/pass through the middle small-diameter channel. At the same time, punch B acts as the backpressure (P_B) on the billet to restore it to its original form. For this purpose, this punch is located at the beginning of the outlet channel. At this stage, P_B is smaller than P_A. Therefore, according to Figure 6.1b, the billet gradually enters the outlet channel. In the second step, P_B, unlike before, is adjusted to be more than P_A, directing the billet to the upper channel through the smaller channel, and punch A is applied as a backpressure punch to restore the billet to its initial dimensions (Figure 6.1c). This deformation sequence can be repeated frequently till the billet cracks. Ultimately, a relatively large metallic billet is obtained with the desired amount of plastic strain and appropriate mechanical properties. It should be implied that the opposite punch is ejected in order to let out the billet during the final extrusion pass. Also, the generated equivalent plastic strain (ε) in the billet after the n-pass process is calculated with the following relationship in which D and d are the die channel diameter in the outer and the small middle region, respectively.

$$\varepsilon = 4n\ln(D/d)$$

6.2 MODIFICATIONS OF CEC-BASED TECHNIQUES

Despite the advantages of the CEC method leading to substantial improvement of the properties, non-uniform plastic strain distribution in the billet due to friction and the need to use a specific and precise backpressure punch are major limitations. The other limitation is the ratio between height to diameter to avoid buckling stresses during extrusion. These factors limit the use of CEC as compared to equal channel angle pressing (ECAP) processes, though the processes were introduced almost simultaneously. There is limited research on the CEC method compared to ECAP. However, several CEC-developed methods have been introduced and investigated over the past decade.

6.2.1 Hydrostatic Cyclic Extrusion Compression Process

One of the limitations of the CEC method that has received less attention so far is its inability to produce long billets, restricting its application on an industrial scale. In the conventional method, the frictional forces increase substantially with an increment of billet length, while the deformation forces remain constant. Consequently, the required total processing force increases significantly, and this force employment causes deformation or buckling of the punch. The longer the billet length, the greater the required force, making the process impossible because of the necessity for high-tonnage hydraulic presses. In the CEC, the billets with a length-to-diameter ratio of up to 4 can only be used. Hydrostatic cyclic extrusion compression (HCEC) method has been proposed for producing quite long-length ultra-fine grained (UFG) billets that can be used readily on an industrial scale. In this new method, shown schematically in Figure 6.2, a pressurised fluid encompasses the billet during the process in order to eliminate the frictional forces, thus significantly decreasing the required processing force.

In other words, due to the high hydrostatic pressure of the process, which results in almost zero friction, there is no particular restriction on the dimensions of the cross-sectional area and length of the billets. As exhibited in Figure 6.2b and c, although the fluid fills around the billet and eliminates the frictional forces between the die and sample surfaces, the sample is in direct physical contact with the die in the HCEC MDZ, the same as the conventional CEC method. Also, a seal is embedded in the back of the fluid, which is pushed by punch A to press the billet in the extrusion area, as shown in Figure 6.2d. Meanwhile, punch B, which is located in the die outlet channel, plays the backpressure role. It also supports the billet to restore its original dimensions. Care has to be taken while filling the fluid in the outlet channel and

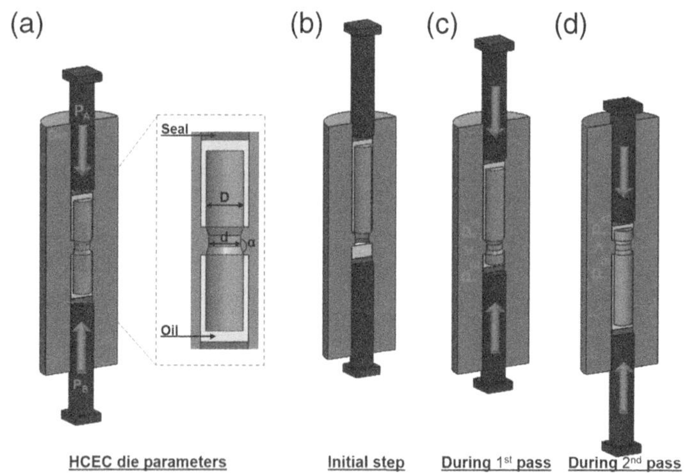

FIGURE 6.2 Schematic representation of the hydrostatic cyclic extrusion compression: (a) the die parameters of the HCEC and (b–d) the steps of the first and second passes in this process.

sealing it. Both punches move downwards because of the low value of P_B compared to P_A. This permits the billet's entire length to pass through the MDZ, thus completing the first pass of the HCEC process. To perform the second pass, punch B pushes the fluid to drive the billet upwards, and P_B, which is greater than P_A, moves upward with the billet. Such a sequence is repeated as many times as needed to increase the amount of plastic strain. The total plastic strain accumulated in the sample after the n-pass of the HCEC process is similar to the conventional CEC method.

6.2.2 Cyclic Expansion-Extrusion

Cyclic expansion-extrusion (CEE) is one of the most famous methods developed in CEC. In general, this process consists of two deformation steps. In the first one, called the expansion step, the billet expands in diameter. Then, the extrusion process is performed on the expanded billet to restore its original dimension. Figure 6.3 presents the sequential steps of the CEE process.

As shown in Figure 6.3, at the beginning of the process, an auxiliary (blocking) punch is placed at the beginning of the die outlet channel to prevent the billet from leaving the MDZ. As a result, the primary punch pressure causes the material to flow radially to fill the die cavity, and the billet takes the barrel shape (Figure 6.3b). Then, the auxiliary punch that blocks the outlet channel is removed, and the primary punch provides the necessary pressure for the extrusion of the material (Figure 6.3c). To perform the second pass, the primary punch is placed

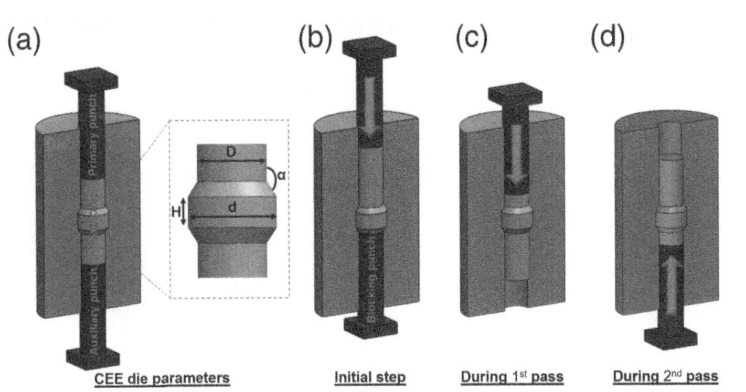

FIGURE 6.3 Schematic representation of the cyclic expansion-extrusion process: (a) the significant CEE die parameters and (b–d) the processing conditions of the first and second passes of the CEE process.

in the lower (outlet) channel, and it pushes the material in the opposite direction without the necessity to apply backpressure (Figure 6.3d). Therefore, this process is repeated many times without removing the sample from the die. There is no need for an auxiliary punch in this CEC-developed method, except at first. The amount of applied plastic strain to the billet after the *n*-pass of the CEE process is

$$\varepsilon = 4n\ln(d/D)$$

Recently, CEE with an asymmetrical extrusion cavity (CEE-AEC) has been introduced and examined, which has a different material flow compared to the conventional CEE process.

6.2.3 Hydrostatic Cyclic Expansion Extrusion

Hydrostatic cyclic expansion extrusion (HCEE) was proposed and patented in 2017 for producing relatively long-length UFG billets. It is similar to HCEC, a pressurised fluid encompasses the billet during the process in order to eliminate the frictional forces, thus significantly decreasing the required processing force. The processing steps are similar to CEE.

6.2.4 Cyclic Extrusion Channel Angular Pressing (CECAP) Process

CECAP is another CEC-developed method based on the combination of CEC and ECAP. Figure 6.4 represents the schematic representation of this procedure. A bulk cylindrical sample stands in the die inlet channel.

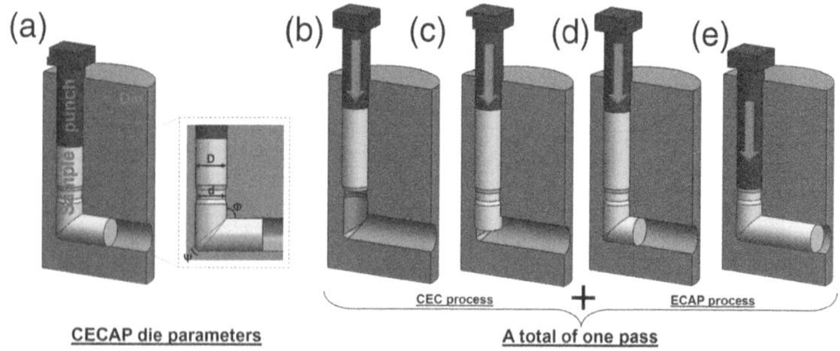

FIGURE 6.4 Schematic representation of the cyclic extrusion channel angular pressing: (a) the significant CECAP die parameters and (b–e) the processing steps of the CECAP process.

First, the initial cylindrical material is extruded to decrease its diameter (Figure 6.4b), and it is compressed after reaching the end of the channel (Figure 6.4c). The billet is then pressed into the angular channel (the conventional ECAP method) to extrude laterally (Figure 6.4d). The angular part of this die actually implements the required backpressure to compress the billet so that the diameter returns to its initial value. In other words, the outlet (horizontal) channel temporarily obstructs the billet from moving and expands it under pressure for returning the billet to its original diameter, which causes more plastic strain fabrication in the billet than in the CEC method. Also, unlike the CEC process, the CECAP method provides the required backpressure for the compression stage without needing a double pressing system. The accumulative strain is due to the sum of the plastic strain in the ECAP and CEC processes.

6.2.5 Cyclic Contraction/Expansion Extrusion

Cyclic contraction/expansion extrusion (CCEE) is another CEC-developed method that includes two consecutive cyclic extrusion compression and cyclic extrusion-expansion processes. Figure 6.5 displays the schematic representation of the CCEE method with the processing details.

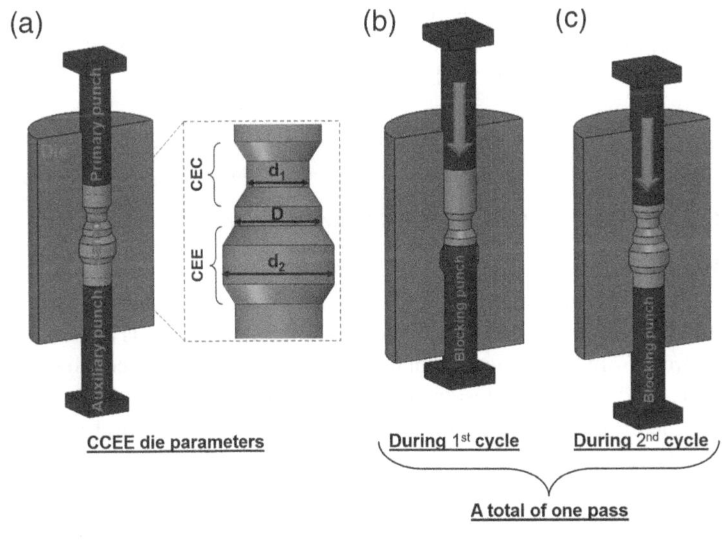

FIGURE 6.5 Schematic representation of the cyclic contraction/expansion extrusion: (a) the significant CCEE parameters and (b & c) the two steps in the initial pass of the CCEE process.

In the first stage of deformation, a cylindrical sample with an initial diameter of D is extruded into a reduced cross-sectional neck with a diameter of d_1. It means that the extrusion process, followed by the billet contraction, occurs at this stage. For this aim, an auxiliary (blocking) punch is placed at the bottom of the deformed billet at the specified location between the two main deformation zones of CEC and CEE inside the die channel, as shown in Figure 6.5b. As a result, the material flows radially and reaches the die wall so that the billet diameter returns to its original value D (Figure 6.5b). Also, the auxiliary punch is applied in the second deformation stage, despite its different location (Figure 6.5c). More precisely, as the primary punch moves further toward the auxiliary counterpart, the billet becomes compressed, and its diameter expands to the diameter of d_2 and then contracts to its original diameter by applying an extrusion process. The auxiliary punch at the beginning of the outlet channel blocks the billet movement in the respective channel. As a result, it is removed from the die channel at the end of the CCEE deformation. Indeed, the expanded billet in the last stage of deformation provides the required backpressure for subsequent passes. An important aspect to be noted is that this process imposes very high plastic strain in one pass compared to the CEC and/or CEE methods. Also, without any need for billet removal from the die, the CCEE process is repeated until the desired amount of plastic strain is obtained. Eventually, the amount of total plastic strain after the n-pass of the CCEE process can be:

$$e = 4n\left\{\mathrm{In}\left(D/d_1 + \mathrm{In}d_2/D\right)\right\}$$

6.2.6 Tube Cyclic Extrusion Compression

While the previous methods are used to process bulk billets, the tube cyclic extrusion compression (TCEC) method (Figure 6.6) is used for tubular samples. In the TCEC method, a tubular sample is constrained entirely from both inner and outer walls by placing it between the mandrel and die. It is also fastened with two endcaps.

The attached short, thicker cross-section to the mandrel provides an MDZ between the mandrel and die for the processing samples, as shown in Figure 6.6a. The enclosed tube can be processed by pushing the mandrel downward until the entire cross-section of the sample passes through the neck-shaped region. The TCEC is a cyclic process in which the tube cross-section is initially extruded from the neck-shaped region and then compressed to its primary dimensions.

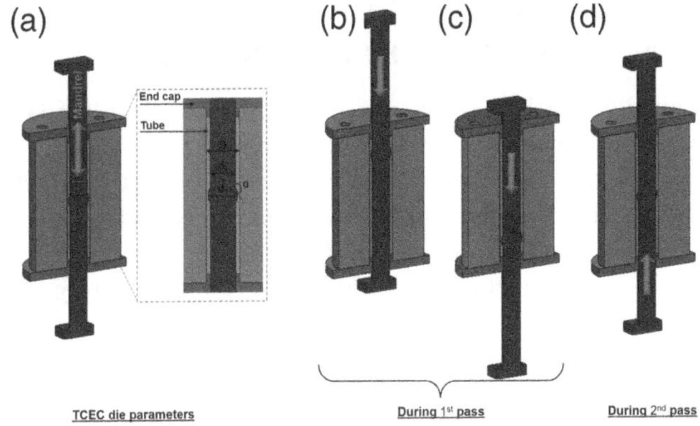

FIGURE 6.6 Schematic representation of the tube cyclic extrusion compression: (a) the significant TCEC die parameters and (b–d) the processing steps in the first and second passes of the TCEC method.

Like previous methods, it is possible to repeat this process by changing the pressure direction (or by rotating the die upside down after each pass due to the symmetrical design of the die setup) between consecutive passes in order to impose intense plastic strains on the tubular sample (Figure 6.6d). Also, the total plastic strain after the n-pass of the TCEC method can be determined from the following equation.

$$\varepsilon = 2n\,\{\ln\left[\left(D^2 - d_2^{\,2}\right)/\left(D^2 - d_1^{\,2}\right)\right] + \left(4/\sqrt{3}\right)\cot(\alpha/2)\}$$

6.2.7 Hydrostatic Tube Cyclic Expansion Extrusion

Hydrostatic tube cyclic expansion extrusion (HTCEE) is a combination of the tube cyclic expansion extrusion (TCEE) method and hydrostatic fluid pressure in which there is almost no friction between the tubular sample and die. The processing setup includes a die, primary punch, moveable mandrel, auxiliary (blocking) punch, oil, and seal. The operational procedure is the same as that of TCEE. One of the beneficial features of this process is that there is no metal-to-metal contact between the tubular sample and the die; hence, the friction is relatively negligible and also the surface roughness is lower. Also, the load and its amount are independent of the sample length due to the nature of this process.

6.3 SOLID-STATE RECYCLING OF ALLOYS AND COMPOSITES BY CEC PROCESS

Traditionally, machining has been one of the most fundamental processes in manufacturing various components, especially in the aerospace and automotive industries. In addition to the direct costs associated with the machining operations, a considerable amount of material is wasted in the form of machining chips. For instance, about 80% of the metal can be lost by machining during the fabrication of medical and aerospace equipment. Therefore, recycling machine chips is a must, and the products obtained by recycling will be cost-effective. There are two basic processes for recycling metal scraps, including the conventional remelting method and the recently developed solid-state recycling process. In the case of remelting recycling, despite its long history of employment, it faces significant problems. The first limitation is the relatively high amount of metal loss due to the essential chemical reaction and the higher specific surface area of the metal chips. The second problem is the production of toxic gases due to the combustion of oil emulsions adhering to scrap chips, and the last issue is related to high energy consumption and recycling costs. Therefore, considering recycling efficiency, energy consumption, cost, and environmental impacts, the solid-state recycling process is more suitable for scrap chips, which can directly convert the scraps into products or semi-finished products with superior mechanical properties without remelting. In this regard, during the last decade, several studies on the solid-state recycling of chips have been performed using traditional and advanced plastic deformation methods, such as hot extrusion, mechanical milling and extrusion, CEC, and ECAP.

Cyclic extrusion compression backpressure (CECBP) technique for the direct-recycling process is represented in Figure 6.7.

Initially, the die cavity is filled with chips. As shown in Figure 6.7c, a whole cycle is initialised and finalised in the compression process. The next stage after the compression process is extrusion using a spring as the backpressure, as shown in Figure 6.7d. Eventually, the sample is ejected when the extrusion process is completed (Figure 6.7e). At the end of the cycle, the sample is turned back to the compression mode with the same processing conditions such as holding time and pressure. Figure 6.7f illustrates the ejection process after completing the CEC process. The compacted aluminium chips sample is shown in Figure 6.7g.

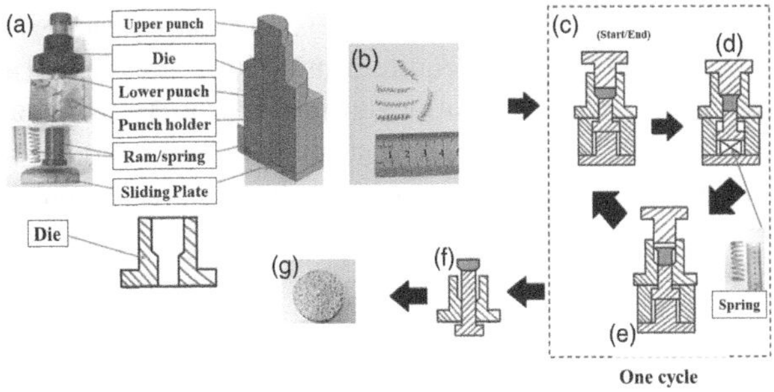

FIGURE 6.7 (a) Cyclic extrusion compression backpressure (CECBP) die setup, (b–g) the procedure and stages of the CECBP process: (b) prepared chips, (c) compression process, (d) extrusion process, (e) ejection process, (f) final ejection process and (g) compacted product.

6.4 ADVANTAGES AND LIMITATIONS

The CEC process has unique benefits over other SPD methods. One of the main advantages of this process is the ability to impose a large deformation over a pass, which is proportional to the D/d ratio. Other advantages include intense grain refinement and high hydrostatic pressure, which prevents billet failure due to the three-dimensional compressive stress condition, enabling the processing of many hard-to-deform materials such as hexagonal-close-packed (hcp) metals and alloys at relatively low temperatures. Backpressure utilisation is another benefit of the CEC process that helps to increase the hydrostatic pressure and causes superior mechanical behaviour of the materials. Furthermore, a backpressure punch increases the number of passes by facilitating the crack closure during processing. Also, repeating this plastic deformation in order to implement high plastic strains does not require removing the billet at intervals between consecutive passes. Therefore, large plastic strain can be attained without removing it from the die during the successive passes, which is another benefit of this method compared to other SPD techniques such as equal channel angular pressing. Finally, sample processing and equipment manufacturing are simple, easy to install, and inexpensive. Generally, CEC-based methods demonstrated acceptable or even better results compared to other SPD-based methods.

Regardless of the many advantages that the CEC method has, unfortunately, it also has some limitations and disadvantages such as its inability to apply uniform strain distribution that causes non-uniform microstructure and resultant non-uniform properties; its complicated die structure which usually consists of several parts leading to increment of frictional forces and requirement for high-stress application; its need for back pressure application during the process, the requirement for successive processing passes to attain the desired strain as well as its limitations on the geometry, size, and shape of the samples, etc.

7 Multi-Axial Forging

7.1 PRINCIPLE AND PROCESS

The working principle of multi-axial forging/multi-directional forging is shown in Figure 7.1. In the multi-axial forging (MAF) process, the material must withstand very high pressure and large strain without any accompanying change in the cross-sectional dimensions of the sample. By using the same mould to process the material multiple times, high strain can be imparted to the sample. This leads to the accumulation of plastic strain in the material. Materials with FCC and HCP crystal structures are processed by MAF to obtain ultra-fine grains.

The major process parameters associated with MAF are deformation temperature, strain per pass, deformation mode and strain rate. The selection of appropriate deformation temperature and deformation strain leads to the necessary grain refinement. MAF with high processing temperature can produce a UFG structure even in brittle materials such as Magnesium (Mg). Also, the high processing temperature reduces crack formation, thereby improving the workability of the material. But at high temperatures, grain growth occurs, and the final grain size increases. MAF at lower processing temperature can contribute to simultaneous improvement in strength as well as ductility as a result of formation of fine grain structure. So, it is necessary to

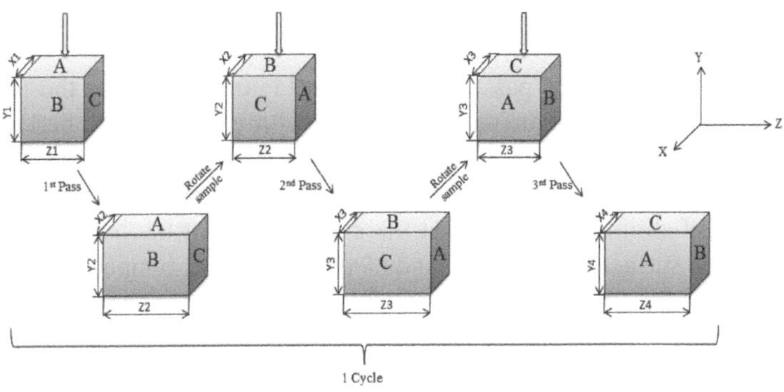

FIGURE 7.1 The three passes of MAF in three different directions X, Y and Z, respectively.

DOI: 10.1201/9781003596110-7

optimise the process parameters in order to obtain a UFG microstructure by MAF. The optimisation of process parameters is also necessary to eliminate the inhomogeneity of the material property in the material processed by MAF.

7.2 PROCESS PARAMETERS

The main process parameters in MAF are the strain and its distribution as well as temperature and its distribution. The way in which the strain and temperature are distributed in the billet is responsible for the amount of grain refinement. The inhomogeneous distribution of strain and temperature causes differences in properties between core and surface, which is measured in terms of the inhomogeneity factor (IF).

7.2.1 Strain and Its Distribution

The true strain in each pass is similar to conventional forging. The true strain (ε_1) in each pass in one direction is given as

$$\varepsilon_1 = \text{In}\left(h_f / h_0\right)$$

where h_f and h_0 are the final and initial dimensions, respectively, of the sample in the direction of the compression. So

$\varepsilon_2 = \varepsilon_3 = -\varepsilon_1/2$ (considering material to be isotropic). If the material is anisotropic, ε_2, ε_3 need to be determined using strain gauges.

Thus, the effective strain (εe_{ff}) after each pass is

$$\varepsilon e_{ff} = \left[2/3\left(\varepsilon_1^{\,2} + \varepsilon_2^{\,2} + \varepsilon_3^{\,2}\right)\right]^{1/2}$$

Each cycle has three passes as shown in Figure 7.1, and so the total effective strain accumulated after one cycle (three passes) will be 3 εe_{ff}.

The strain thus calculated will not be distributed uniformly since the magnitude varies along its dimensions. Figure 7.2 represents the distribution of strain accumulated at the cross section of the half distance from the surface during MAF at different processing temperatures.

The strain distribution forms an X shape in the area under observation. As the number of passes increases, uniformity increases. Higher uniformity is observed at the end of six passes at all temperatures. The accumulation of strain increases the dislocation density of the material. At higher temperatures, the recovery occurs through annihilation and rearrangement of dislocation. Increase in dislocation density and

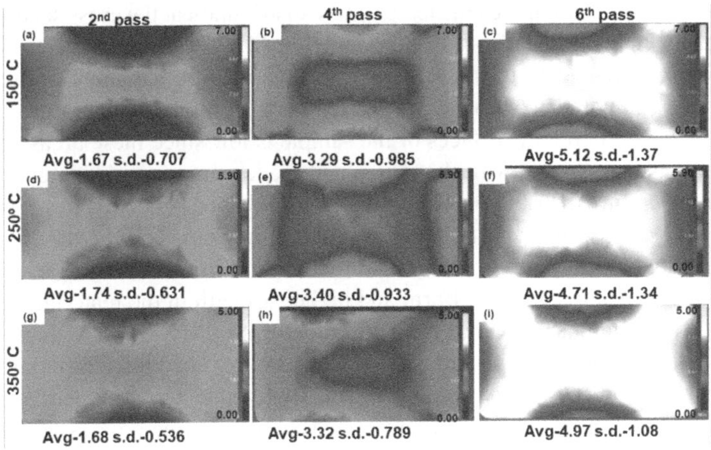

FIGURE 7.2 Strain distribution for MAF performed at different temperatures.

annihilation governs the grain refinement, and so, generally, fine grain also starts forming at the centre in the initial MAF passes, and gradually an X-shaped fine-grained zone forms in subsequent passes. Strain distribution and grain size distribution will be almost the same.

7.2.2 Temperature and its Distribution

The MAF processing temperature significantly affects the microstructure evolution and the mechanical properties. MAF based on recrystallisation temperature of alloy, like conventional metal forming processes, can be divided into cold MAF, performed at room temperature, warm MAF performed at a temperature below recrystallisation, and hot MAF performed at a temperature above recrystallisation temperature. Cold MAF is simpler than either warm or hot MAF, and cold MAF also facilitates the development of dislocation substructures inside deformed grains even when the same amount of deformation is applied, promoting grain fragmentation. On the other hand, although there are differences depending on the alloy composition when using warm and hot MAF, the dynamic softening behaviour and strain-induced transformation if any, affect the grain refinement. Hot MAF is effective with an increasing number of cycles, but annealing heat treatment is required to secure the mechanical properties, and the final microstructure has a bimodal nature. Besides this, MAF is also carried out at low temperatures, i.e., cryo temperatures for the metals and alloys exhibiting sufficient ductility at cryo temperature.

Figure 7.3 compares the temperature distribution at the cross section of the half distance from the surface after MAF at different temperatures. It is evident that MAF at higher temperature has more uniformity compared to MAF at lower temperatures. The temperature at the top and bottom surfaces of the sample is low since these areas are in constant contact with the punch and die kept at room temperature. Temperature distribution also follows the X-shaped pattern as in strain.

In spite of the fact that MAF at lower temperatures has better accumulated strain with more uniform distribution in the material, which can ultimately lead to uniform and higher strength in the material, but the lower temperature may promote cracks if the material does not have sufficient ductility at the processing temperature. Therefore, lower temperature processing may not be practically feasible. On the other hand, a higher processing temperature can ensure defect-free processing and more uniformity in temperature distribution leading to less strain accumulation. So, optimisation of processing temperature is required in order to ensure the feasibility of the process without any failures and a homogenous distribution of strain and temperature. It is necessary to start the processing at a higher temperature and then slowly decrease to a lower value. The gradual decrease in temperature will cause more strain accumulation, which in turn will result in grain refinement. Generally, the initial passes are carried out in high temperatures and subsequent passes at lower temperatures. This will ensure low forging forces during initial high temperature processing and lower chances of failure of the material at the lower temperature processing.

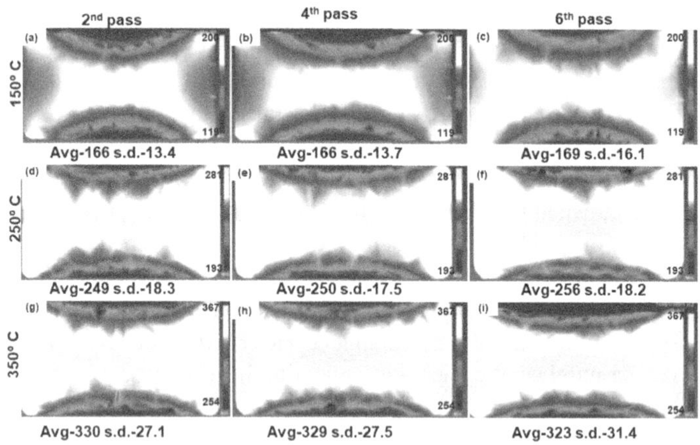

FIGURE 7.3 Temperature distribution for MAF performed at different temperatures.

7.2.3 Inhomogeneity Factor

Inhomogeneity factor (IF) reveals the homogeneity achieved in the processed billets. This can be quantitatively determined using the following equation:

$$\text{Inhomogeneity factor}(\text{IF}) = \frac{\sqrt{\sum_{i=1}^{n}\left(H_i - \bar{H}\right)^2 / (n-1)}}{\bar{H}}$$

where n is the number of hardness measurement in each sample, H_i and H are the ith hardness measurement and mean hardness, respectively. Generally, for many metallic systems, IF will improve after six passes.

The distribution of hardness level in 1-cycle is relatively highest in the centre region and lowest in the corner zone parallel to the three axes. In the 2-cycles specimen, the portion with a relatively high level of hardness is distributed in the diagonal direction and includes the centre and face-1 regions. This result is similar to measurements of the 1-cycle specimen, and both conditions showed a relatively low level of hardness in the edge region parallel to the three axes. The 3-cycles specimen, however, had relatively high levels of hardness values in the centre region and surrounding zone, and the maximum hardness values were measured in the centre region. In the 4-cycle specimen, the highest hardness value is found in the centre region. Figure 7.4 displays a comparison of the average hardness values for each specimen, which includes the hardness values for the centre regions. The average hardness values increased rapidly in the 1-cycle specimen, and no significant change occurred as the number of cycles increased. This behaviour seems to be closely related to the softening behaviour that occurs during deformation.

The properties of the deformed material are a function of strain rate and the processing temperature and can be represented by Zener-Holloman parameter.

$$Z = \dot{\varepsilon} \exp\left(\frac{Q}{RT}\right)$$

where ε and T are the strain rate and the processing temperature, respectively; Q is the activation energy, and R is the gas constant. As per the equation, high strain rate and low processing temperature will result in higher Z, which may result in improvement in grain refinement. Thus,

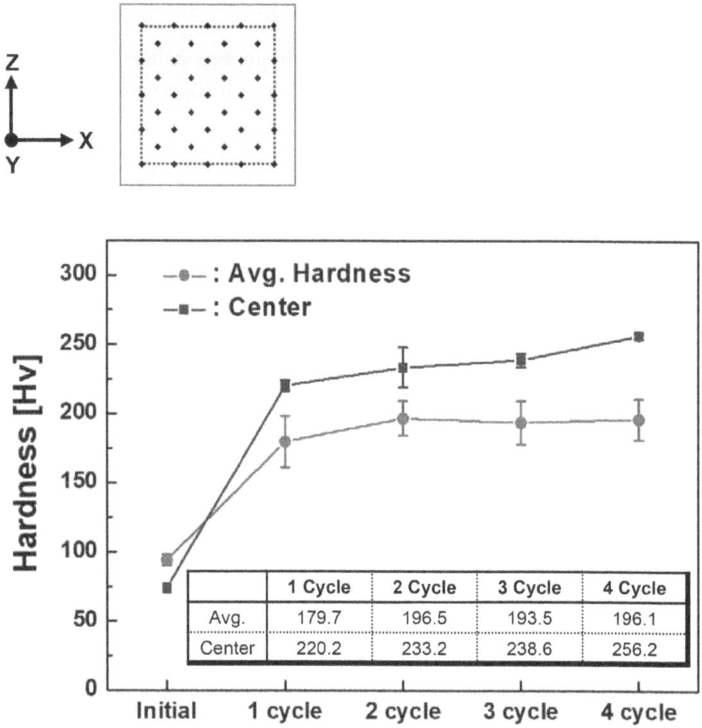

FIGURE 7.4 Vickers hardness of the centre region and average hardness of specimens deformed by different numbers of cycles.

a decreasing processing temperature with increasing MAF passes (i.e. increasing strain) is advantageous to achieve a higher degree of grain refinement.

7.3 MULTI-AXIAL DIAGONAL FORGING

Regardless of the process temperature, materials manufactured by MAF have some inherent problems. Multi-axial diagonal forging (MADF) has been proposed to overcome this problem (Figure 7.5). This process consists of plane forging (PF) and return-plane forging (R–PF), diagonal forging (DF), and return-diagonal forging (R-DF). The six passes of the first cycle consist of PF and R–PF, which is followed by six passes that involve DF and R-DF. In the PF process, a cube-shaped workpiece is placed in a fixed channel die, and pressure is applied using a flat punch in a direction perpendicular to the surface of the workpiece.

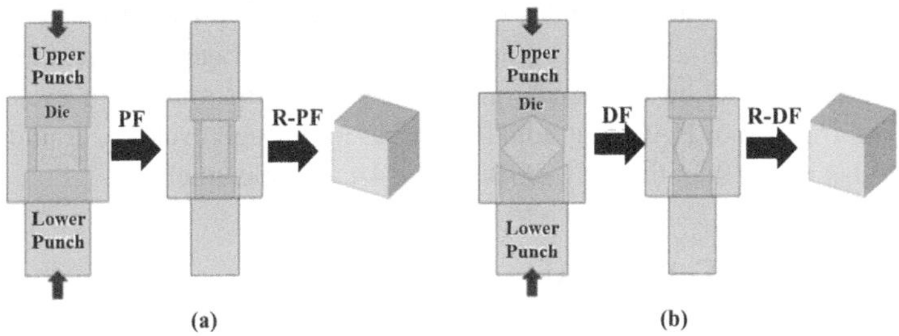

FIGURE 7.5 Schematic diagrams of typical forging operations with top and bottom punches, a die and a workpiece: (a) PF, R-PF and (b) DF, R-DF.

In the R–PF process, the elongated workpiece deformed by the PF process is rotated and compressed to return to the initial cube shape using a flat punch. In the DF process, the cube-shaped workpiece is cut at 45° and inserted into a fixed channel die, and then a curved punch presses the workpiece diagonally. In the R–PF process, the workpiece is rotated 90°, inserted into a fixed channel die, and compressed using a flat punch to return it to its initial cube shape. Theoretical studies using FEA have confirmed that DF and R-DF induce a uniform distribution of effective strain in the deformed specimens compared with PF and R–PF.

Compared with MADF with PF-RPF, MADF with DR-RDF requires less frequent mould replacement and induces relatively high levels of hydrostatic pressure inside the material that suppresses the generation of forging defects even when the number of cycles increases.

The average grain size decreased sharply in MADF compared with the initial specimen after 1 cycle and showed a tendency to decrease continuously with the increased number of cycles. A decrease in grain size occurring in a deformed specimen while the number of cycles is increased is directly related to the extensive development of the deformation-induced boundaries (DIBs). In addition, by comparing the grain size developed in each region, it was possible to grasp the heterogeneity (inhomogeneity factor) of the microstructure. In the case of the 1-cycle specimen, the deviation of the grain sizes measured in each region and the inhomogeneity factor are largest, and both grain size and inhomogeneity factor gradually decrease as the number of cycles increases. Most of the fine grains are developed in the centre region regardless of the number of cycles.

After 1 cycle, the average values for misorientation angles were decreased by half compared with the initial specimen, but the averages continued to increase after 2, 3, and 4 cycles. In all materials, the differences in the misorientation angles between each region are similar, and all misorientation angles of the centre regions proved to have the largest value under each cycle. As with the average misorientation angle, the fraction of HAGBs also depended on the number of cycles. The fraction of HAGBs gradually increased through cycles 1, 2, and 3 and is larger than that of LAGBs in the 4-cycle specimen.

7.4 CLOSED DIE MAF

MAF can be executed with an open channel die or a closed channel die. However, it is a fact that more dimensional accuracy in the samples could be obtained using a closed channel die. In this closed die MAF, the sample is pressed to a fixed strain in a specially designed die, removed and rotated to 90°, reinserted in the die, and the same amount of strain is applied again by pressing and so on.

To carry out MAF at room temperature, a closed MAF die (Figure 7.6) is used so that the samples can be multiaxially forged in three orthogonal directions (X, Y, and Z-axes). It is to be observed that the width of the sample remains constant, and 50% normal strain can be imposed on the material during each pass. Plane strain conditions can easily be maintained. The effective strain (ε_{eff}) imposed on the material during each cycle of MAF under plane strain condition can be determined using the equation

$$\varepsilon_{eff}^{\text{One cycle}} = N\left[\frac{2}{9}\left\{(\varepsilon_{11})^2+(\varepsilon_{22})^2+(\varepsilon_{33})^2\right\}\right]^{0.5} = N\frac{2\sqrt{3}}{3}\ln\frac{h_0}{h}$$

where N is the number of passes carried out in the sample, which needs three passes to complete one cycle of MAF. ε_{11}, ε_{22} and ε_{33} are axial strains in three orthogonal directions; "h_0" and "h" are initial and final

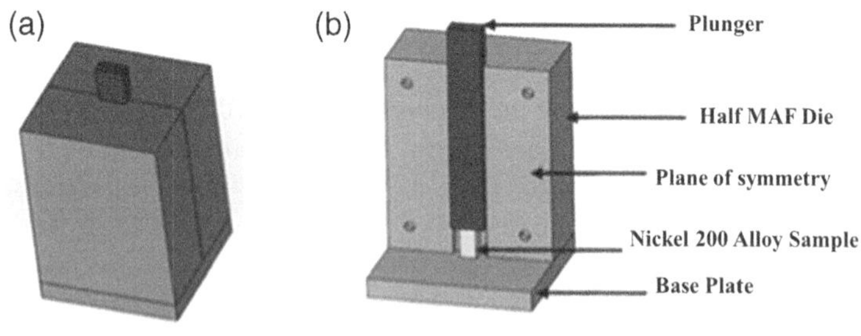

FIGURE 7.6 Closed die MAF process (a) MAF-die assembly (b) illustration of MAF execution.

heights of the samples. During MAF of each pass, the sample has to be rotated by 90° and then put into the die for forging.

7.5 ADVANTAGES AND LIMITATIONS

Multi-axial forging is the right process that balances cost, speed, and quality among the SPD processes. The advantage of MAF is that the applied induced strains can be increased up to the required levels, and rotation of sample after each pass by 90° helps in maintaining the initial shape. While the advantages of MAF lie in its simple tool design and ability to process bulk materials, the main limitation is the inhomogeneity due to the localised deformation along the direction of shear, resulting in inhomogeneous microstructure across the cross section at initial passes. However, increasing the number of MAF passes may partially help to solve the problem, but an arbitrary increase in the number of passes may lead to a redundant increase in manufacturing cost and time. The other major limitation is effectively controlling the temperature. Uneven distribution can cause problems such as hot tears or weak spots in the material. This is especially true when working with large parts where the heat may not spread evenly throughout the component. Certain alloys, like nickel-based superalloys, require even more careful temperature management.

Most investigations of MAF have concentrated on using this procedure to process aluminium alloys, copper alloys, ferroalloys and titanium alloys, while little research has been conducted on other alloys.

8 Accumulative Roll Bonding

8.1 PRINCIPLE AND PROCESS

Accumulative roll bonding (ARB) is a multi-step roll bonding, using a simple conventional rolling mill, in which, firstly, the mating surfaces are cleaned to remove the sediment layers, followed by wire brushing to enhance the bond strength between the materials. Automated cylindrical brushes instead of mechanical brushes are also used for cleaning the surface. The sheets are analysed for different roughness parameters and the surface topography. Next, the layers of the materials are joined together and passed through the rolling mill, and large plastic deformation is applied to cause material deformation to provide necessary bonding and grain refinement. The material, which is rolled and cut into two pieces, and these pieces are accumulated again according to the initial dimensions, and the whole procedure is repeated limitlessly. Figure 8.1 depicts a schematic diagram of the ARB process. The same method can be modified to process composites by including an extra step, i.e., dispersing secondary phase particles after surface preparation. The bond strength and the number of ARB passes are the crucial factors that decide the quality of ARB. The heat generated between the rolls due to

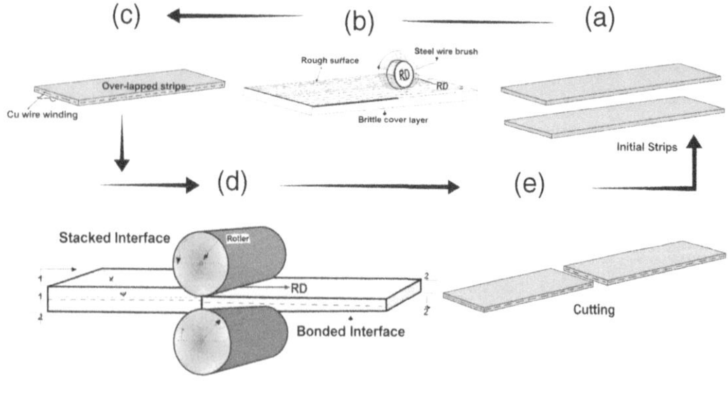

FIGURE 8.1 Schematic illustration of ARB process.

DOI: 10.1201/9781003596110-8

the presence of roll pressure results in good bonding. The ARB process is usually performed at elevated temperatures or below recrystallisation temperatures to achieve strong bonding and grain refinement.

In the ARB process, the material type and extent play a major role, and the materials considered in the ARB process are classified into three major classes i.e. (i) similar materials (ii) dissimilar materials and (iii) similar/dissimilar materials with reinforcements. The key factor in ARB is the bonding between the plates to maintain the dimensions. It is important to understand the bonding mechanism.

8.1.1 Bonding Mechanisms and Material Parameters

Mechanisms of solid-state bonding between the materials during roll bonding are categorised mainly into energy barrier, film theory, recrystallisation and diffusion bonding. Film theory explains that the bonding between the two solid-state materials is generated when the surfaces are deformed to a particular value and bonding is created when the bonding prerequisite overcomes the energy barrier (diffusion) of the material, even at low temperatures. There is another theory that the bonding is due to the crystal misorientations at the contact and not from diffusion and recrystallisation. Generally, the interface bonding process is explained by the Bay mechanism and the diffusion theory during the SPD. The schematic diagram is shown in Figure 8.2. Although the surface treatment before rolling can grind off the oxide film on the surface of the raw plates, once in contact with air, they will get oxidised again. The oxide film can be crushed during the ARB process, exposing the surface to fresh metals. The pressure in the normal direction (ND) causes the active surfaces of the two metals to contact each other. The nature of this process is to bring the atoms on the surface of the two metals close together and produce a large number of binding sites through atomic adsorption and diffusion. With the increase of cycles of ARB, the fresh metal gets closer and closer, the atomic diffusion is

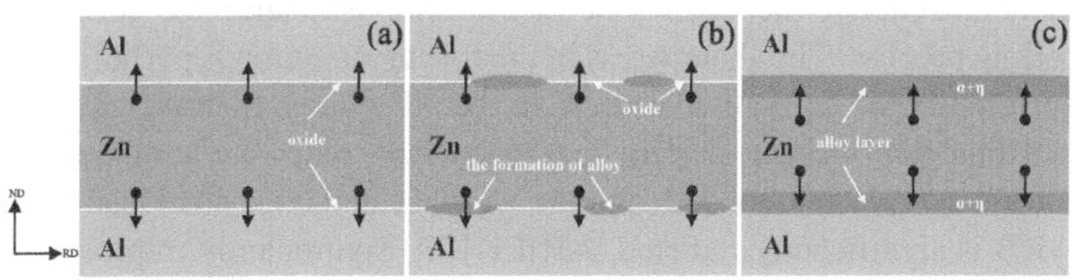

FIGURE 8.2 Schematic diagram of the diffusion of atoms after the contact of Al and Zn layers and the preliminary formation of Zn–Al alloy layers.

more and more sufficient, and the interface is more and more closely combined. This bonding mechanism is also commonly referred to as the crack bonding mechanism. However, the exact bonding mechanism is likely to vary with different metallic systems.

8.2 MODIFICATIONS OF ARB TECHNIQUES

Some variants of the ARB method have been introduced to further enhance the quality and properties of samples compared to the conventional ARB method. Some of these ARB-based variants include hard plate accumulative roll bonding (ARB (HPR)), cross accumulative roll bonding (CARB), ARB followed by cryo rolling, accumulative press-bonding (APB), repetitive press roll bonding (RPRB), and continual annealing and press-bonding (CAPB).

8.2.1 Hard Plate Accumulative Roll Bonding

At present, the processing methods for preparing composite laminates include the extrusion compounding method, explosive compounding method, high-pressure torsion method, Friction stir welding, diffusion bonding, etc. Still, many problems exist, such as high process costs, complex processes, and low production efficiency. ARB is widely used because it is the most productive method to prepare ultra-fine grained composite materials. If it is used as a strengthening means of rolling composite, large deformation broken grains can be obtained through accumulated strain, forming large angle grain boundaries, and refining grains can significantly improve the strength of the material.

In the traditional ARB processing of heterogeneous composite laminates, due to the excessive strain caused by ARB, many problems, such as reduction of plasticity and cracks on the edge of composite plates, cannot be avoided while increasing the strength. Hard plate accumulative roll bonding (ARB (HPR)) is employed to overcome such issues for developing composites such as Mg/Al, Zn/Al, etc. The addition of hard plate can change the stress state of the composite laminate during the rolling process. The shear stress in the rolling direction is partially converted into the compressive stress in the ND, which then increases the bonding pressure between the heterogeneous composite laminates; the interface is closely bonded, and the warpage and edge crack caused by ARB is significantly reduced. ARB (HPR) significantly improves the tensile strength and plasticity of the composite laminate as compared to traditional ARB. The process principle of hard-plate accumulative roll bonding (ARB (HPR)) is shown in Figure 8.3.

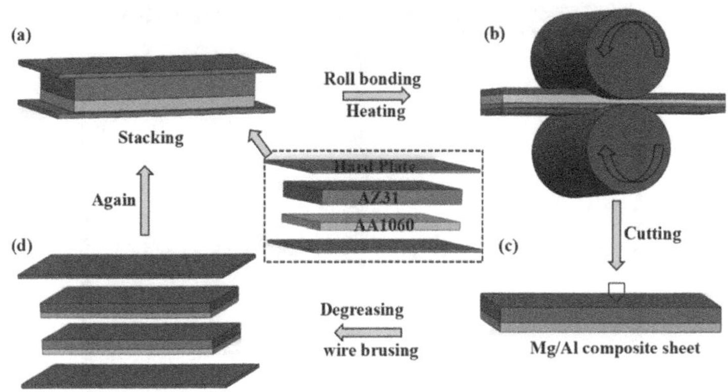

FIGURE 8.3 Process principle of hard plate accumulative roll bonding of composite plate.

Before rolling, the plates need to be surface-treated. First, the plates are polished to expose a new surface, cleaned with acetone solution, and then the surface of the plate is cleaned with alcohol and air-dried naturally. Then, two surface-treated plates are stacked on top of each other. To prevent the relative sliding of the two initial plates, either the sheets are riveted or fixed with thin iron wires. When the hard plate rolls, a layer of high-temperature release agent (boron nitride) should be sprayed on the inner surface of the hard plate to prevent the composite laminate and the hard plate from sticking together, becoming unable to be separated during the hot rolling process. For hard plate rolling (HPR), hard plates are added to the upper and lower surfaces of the laminates as shown in Figure 8.3. After preheating, if necessary, the fixed end of the plate is sent to the rolling mill for rolling, and the reduction is adjusted to 50%. After ARB HPR, the plate is cut into two plates of the same area to repeat the rolling process. In the traditional accumulative rolling bonding (ARB(TR)) of composite plate, there is obvious warpage at both ends of the composite plates, and the plate has subtle edge cracks as shown in Figure 8.4. On the contrary, there is no crack at the edge of composite plates fabricated through hard plate rolling bonding (ARB(HPR)), and the binding is tight.

The addition of hard plates avoids direct contact between the plate and the rolls, and the upper and lower hard plates are slightly larger than the composite plate and play a protective role, thereby preventing the composite plate from shattering. The hard-plate accumulative roll bonding inhibits the generation of edge cracks, enhances the bonding, and uniformity of the composite plate deformation, and plays a specific

FIGURE 8.4 Plates rolled by ARB and ARB (HPR).

role in straightening. Apart from this, in the case of asymmetric rolling, where the upper and lower plates are different materials, this leads to a difference in the friction index of the upper and lower surfaces. During rolling, the forces on the composite plate will be different and so, the composite plates cannot be tightly connected and offset to both sides. Adding a hard plate solves this problem very well.

The addition of hard plates changes the interface bonding compared to traditional rolling, and also, both yield stress and elongation are improved in hard plate accumulative roll bonding. In addition, as the interface remains straight, and the composite layer maintains good continuity, so the tensile strength and elongation are high. In general, under the combined action of adding hard plates to ARB, the microstructure and interface bonding of the matrix plate can be changed, and the overall mechanical properties of the composite plate can be further improved.

8.2.2 Cross Accumulative Roll Bonding

Though ARB increases the strength, the plasticity is significantly decreased. It has been a long-term goal to achieve both high strength and good ductility. In the ARB process, after each cycle of rolling, if the plate is rotated by 90° around the ND, then the process is called as cross accumulative roll bonding (CARB), as shown in Figure 8.5.

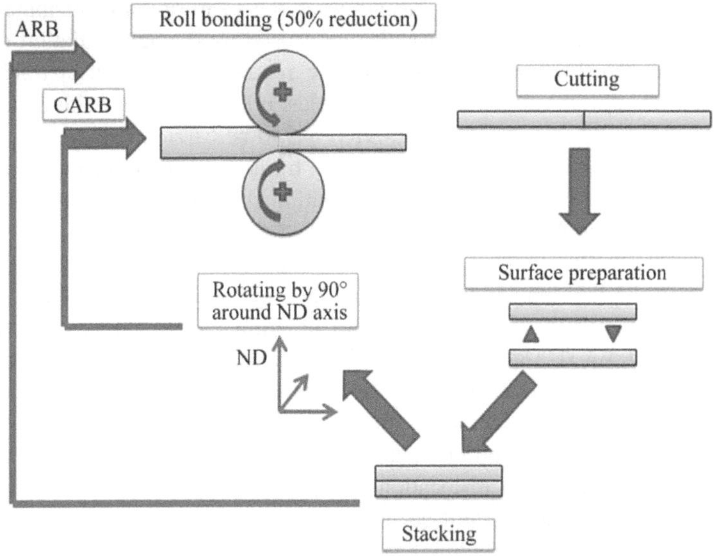

FIGURE 8.5 Schematic illustration of the cross-accumulated roll bonding process.

Materials processed by CARB, with finer and more uniform grains, have better overall performance than those processed by ARB.

Grain size distribution is not uniform in ARB due to the non-uniformity of the rolling deformation, while after CARB many large grains are refined into small grains, and the grains are smaller and more uniform. During the CARB process, the plate is continuously rotated by 90° around the rolling direction and the layer boundary of the plate in the rolling channel is also alternatively parallel or perpendicular in different rolling cycles. This greatly reduces the length-to-diameter ratio between the connection boundary spacing and the lamella boundary spacing, improving the grain refinement efficiency and thereby enhancing the grain refinement strengthening effect. Therefore, the change in rolling path makes the CARB material stronger and better than ARB. During the CARB process, the cross-rolling behaviour causes the sheet to have a weak texture. The weakening of the substrate texture is very conducive to improving the ductility of the material because the slip is easily activated.

8.2.3 ARB Followed by Cryo Rolling

Although ARB, a severe plastic deformation process, is more effective in refining grains than conventional rolling, the grain refinement rate tends to saturate as strain increases. This saturation is attributed to the dynamic recovery growth rate caused by the accumulation of plastic deformation, and the dynamic recovery rate is inseparable from the deformation conditions. Besides, in high melting point materials, to achieve good bonding, ARB has to be carried out at slightly elevated temperatures. Therefore, the limited grain refinement can be overcome by changing the deformation conditions to suppress the dynamic recovery.

Cryo rolling (CR) is a simple low-temperature rolling process in which the low temperature is maintained by liquid nitrogen. CR can increase both the rate and the limit of grain refinement, leading to a finer grain size and higher strength. The cryo-rolled sample has a higher dislocation density than the ARB samples. This is because the dynamic recovery/recrystallisation is suppressed in the CR process, leading to an increase in the number of dislocations and a decrease in the spacing between dislocations. As the deformation continues, the dislocation movement will be subjected to higher resistance, resulting in a rise in the stress required for deformation. Additionally, due to the refinement of grain size by CR, the relative area of grain boundaries increases, and the inhibiting effect of grain boundaries on dislocations is also enhanced. The stress required for further plastic deformation of the material increases, resulting in an increase in strength.

Thus, CR after ARB increases the strength of the materials as compared to ARB. However, in CR, the dimension (thickness) is reduced due to deformation. Little compromise in dimensions has to be tolerated. So, generally, the last pass will be the CR.

8.2.4 Accumulative Press-Bonding

Accumulative press-bonding (APB), as an SPD method with similarities to the ARB technique (but uses a hydraulic press instead of rolling force), was introduced about 10 years ago, with the potential to apply extreme high plastic strains on the material and induce a high amount of structural refinement in order to improve the mechanical properties and produce high-quality metal matrix composites (MMCs).

During the APB process, after accomplishing the usual surface preparation as in ARB-based methods (degreasing followed by brushing), ultrasonicated powder suspension in acetone is sprayed between the two metallic strips with an atomiser. After successful particle deposition and complete acetone evaporation, in which the brushed surfaces of strips are covered homogenously with reinforcement particles, the two strips are placed on top of each other and stacked. The stack strips should be

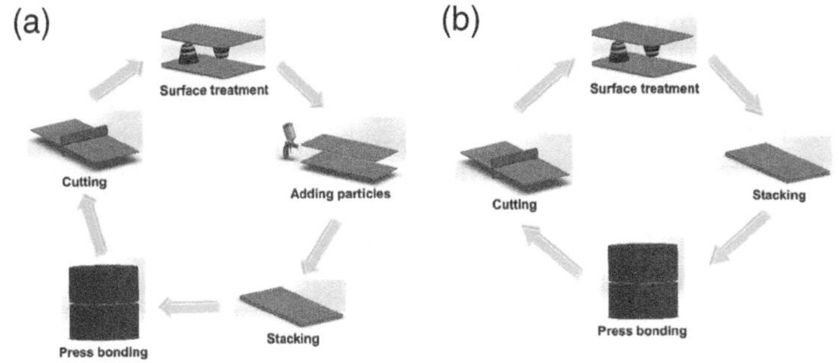

FIGURE 8.6 Schematic representation of accumulative press-bonding (APB) process (a) with (b) without reinforcing particles.

handled cautiously in order to prevent any new contamination. Then, the cold press-bonding procedure is performed on the stacked strips without utilising any lubrication via a hydraulic press machine up to a specific reduction. Then the press-bonded strips are sectioned in half, and the same procedure is repeated usually up to five passes.

The first stage of APB involves specific reinforcement particles (Figure 8.6) into the metal matrix and fabricating various MMCs. While in the second stage, as illustrated in Figure 8.6b, the same procedure is repeated again up to the required number of passes without particle addition. This stage is aimed at achieving an excellent dispersion of reinforcement particles in the matrix. The APB technique is utilised to fabricate the nanostructured composites such as Al/SiCp composite, Al/WCp composite, etc., and the results indicate improved yield strength and ultimate tensile strength as compared to conventional ARB. Furthermore, APB has a great potential in the production of UFG structure so that the gradual generation of nano/ultra-fine structure is achieved in many metallic systems.

Another variant of the ARB method, the CAPB method, is quite similar to the APB method, in which the annealing procedure should be performed after the press-bonding step and before the cutting step in the first and second stages of the APB method.

8.2.5 Repetitive Press Roll Bonding

RPRB is based on the ARB method so as to simplify and facilitate the processing steps of ARB and APB to produce high-quality MMCs. The RPRB process requires many compression cycles, typically between 10 and 16 times, to attain refined grains and a uniform structure with favourable mechanical properties. In this method, after the usual ARB surface preparation steps, reinforcement particles are dispersed between the two strips; then, they are stacked on each other and firmly fixed by Cu wires at both ends. Subsequently, a high pressing machine is used to implement the first processing stage without utilising lubricant. The first stage aims to achieve a firm bonding and is similar to the APB process. However, in the next stage, the bonded strip is pressured in a back-and-forth condition to press the sample evenly on all sides, called repetition. In the third stage, the rolling process is implemented on a sample that has already been pressed and rolled in a back-and-forth condition. Note that the temperature conditions should be selected based on the processed materials (usually above the recrystallisation temperature). Indeed, RPRB is a combination of APB and ARB methods.

8.3 PRODUCTION OF GRADED STRUCTURES AND LAMINATED COMPOSITES

The ARB method can be used to fabricate graded structures with unique properties. This tremendous potential to produce graded structures comes back to its possibility in the fabrication of individual layers during the ARB processing. This offers numerous chances for the invigoration of the layered material stack with varying properties in each layered zone, leading to the generation of tailored and graded structured materials. In addition, ARB is known as an innovative scheme for multi-layer structure fabrication due to its simplicity and economic nature. The ARB method can be utilised to achieve a controlled gradient composition. Air gun spraying is used to spatially distribute reinforcing particles on substrate sheets in order to achieve a graded structure continuously and systematically with reinforcement content along the rolling direction. Figure 8.7 shows the schematic of the experimental setup and the four cycles of the ARB procedure to produce a graded structure. As is seen in Figure 8.7a, an aqueous suspension including desired mass fraction of reinforcement is applied on matrix sheets by air gun spraying with controlled conditions of feed velocity of the nozzle and distance between nozzle and sheet, respectively, leading to the gradation by acceleration and the angle with various particle volumes on the matrix sheets.

This graded and varied distribution of reinforcing particles are spread on two sheets along the longitudinal direction (in the rolling direction) before the ARB implementation. Then, sheets are dried and heat-treated for a pre-determined time and temperature, and two sheets from the

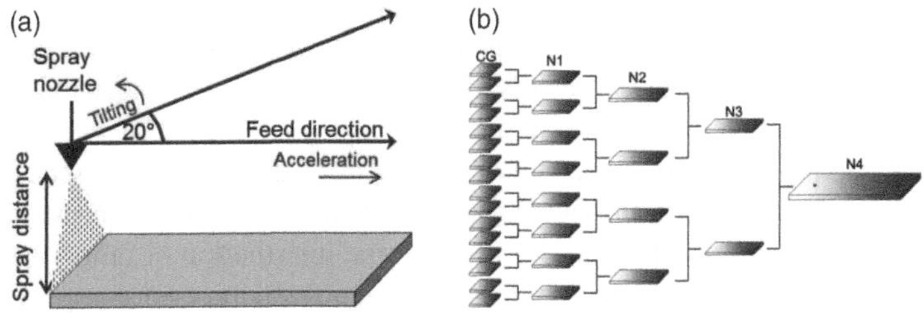

FIGURE 8.7 (a) The schematic of ARB experimental setup with the factors influencing the powder content and (b) ARB processing starting from sixteen small sheets, finally resulting in one sheet after four cycles.

coated surfaces are stacked and roll-bonded up to 50% thickness reduction. As illustrated in Figure 8.7b, in the first cycle, sixteen sheets are processed, leading to eight produced sheets. Subsequently, the same procedure is accomplished on the produced sheets in the second cycle, resulting in the production of four sheets. Finally, after four cycles, this procedure is completed with the final sheet having the required length.

ARB processing has also been utilised to fabricate laminated composites such that dissimilar metals can be sandwiched in alternate layers. The laminated metal composites (LMCs), comprising two or more metals, have been introduced based on their enhanced fracture toughness, improved fatigue life, proper bending characteristics, suitable impact performance, high corrosion and wear resistance, and favourable damping capacity. It was known that laminated composites possess high toughness and strength characteristics, especially along the direction parallel to the layer interface, since they consist of alternatively stacked structures resembling the "brick-and-mortar" condition. Utilisation of the ARB method to manufacture laminated composites can be very beneficial, providing a lightweight structure with superb properties. There are many Al-based metallic laminates such as Al/Mg, Al/Ti, Al/Ti/Al, Al/Ni, Al/Steel, Al/Cu, Al/Ti/Mg, and Al/Cu/Mg laminates produced by ARB. The laminated composite produced by ARB is so powerful that it leads to the production of nanostructures and intensive grain refinement evolutions.

8.4 ADVANTAGES AND LIMITATIONS

ARB is an economical SPD technique and is affordable for different types of metallic materials and alloys. Using the ARB technique, the bonding quality of similar and or/dissimilar materials or materials with reinforcement can be increased, which is not possible using conventional bonding methods. ARB, as a severe plastic deformation process, can improve the microstructure of the matrix alloy and the distribution of reinforcing nanoparticles in the metal matrix after the production of the casting nanocomposite. The first goal is to use ARB to produce a metal matrix composite and the second is to improve the microstructure of the matrix alloy and to improve the distribution of reinforcing particles in the metal matrix after production of the cast composite. The ARB process is also used to remove porosity defects and improve the distribution of reinforcing particles in the casting composite structure as well as applying strain on the matrix alloy of the composite. Also, other structural defects such as poor bond between particles and matrix, and particle-free zones are significantly modified. By increasing the number of cycles in ARB, composites with uniform particle distribution and homogenous microstructure can be achieved.

In ARB, cracks can occur in the metal at the free edges due to free deformation without enough hydrostatic compressive stresses, which can be minimised by hard plate ARB. The other major limitation is the temperature. Low temperatures can result in insufficient bonding and bond strength, while high temperatures can cause grain coarsening and dynamic recrystallisation. With a little sacrifice in dimensions, CR followed by ARB solves this issue. The ARB process is controlled by many parameters, making it relatively expensive. The mechanical properties, especially the hardness of ARB sheets, can fluctuate greatly due to the interface and uneven deformation between sheets. Precise control over deformation and temperature is required.

9 Groove Pressing

9.1 TYPES OF GROOVE PRESSING

Groove pressing (GP) is of two types:

i. Constrained groove pressing (CGP) and
ii. Repetitive corrugation and straightening (RCS).

In CGP, the sheet specimen is placed between the lower and upper dies in a container such that it is tightly constrained. In this condition, when the dies press the sheet specimen, the sheet specimen can move neither in the longitudinal direction (left or right) nor in the transverse direction (out of plane or into the plane). In unconstrained GP or RCS, there is no constraint on the movement of the sheet specimen. In the RCS process, the specimen elongates due to the lack of constraint, and the specimen is resized after each stage of deformation. The wastage of material is approximately zero in the CGP process, while only 84% of the processed sheet length is useable in the RCS process. The RCS process leads to a considerable change in the specimen thickness while the change in the specimen thickness is very insignificant during CGP. The magnitude of average strain imposed in the specimens during CGP is higher when compared to the RCS process. It is also stated that the strain uniformity of the RCS process is higher in the first pass, whereas it is higher in the CGP process for the later passes. Figure 9.1 shows the CGP and RCS die sets.

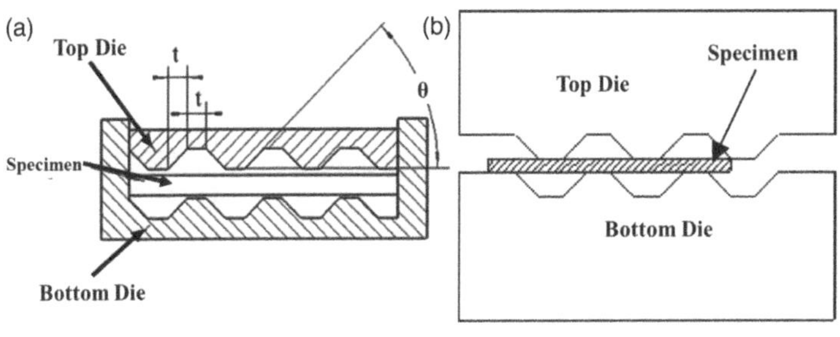

FIGURE 9.1 (a) CGP and (b) RCS die.

DOI: 10.1201/9781003596110-9

The mechanism of grain refinement, principles, methodology, process and geometrical parameters are the same for both cases. In this chapter, CGP and RCS are used as synonyms.

9.2 PRINCIPLE AND PROCESS

The technique of CGP was developed by Shin. In the CGP process, a sheet specimen is subjected to orthogonal shear deformation by repetitive pressing under grooved dies and flat dies alternatively. A schematic diagram of CGP technique is shown in Figure 9.2. Firstly, a pair of asymmetrically grooved dies that are tightly constrained by cylinder wall is prepared. In RCS, there is no constrain from cylinder walls, instead only guiding pins are used. The sample is placed in between the dies and the pressing is performed such that the gap between the upper die and the lower die is the same as that of the sample thickness. This results in pure shear deformation under plane strain condition in the inclined region of the sample (single hatched area in Figure 9.2b). However, the flat region remains undeformed (unhatched area in Figure 9.2b). The equivalent strain will vary depending on the geometrical shape of the die, which will be discussed later.

In the second pressing, the corrugated sample is straightened by means of flat dies (Figure 9.2c). This straightening under the constrained condition ensures that the already deformed region is subjected to reverse shear deformation while the region not deformed remains unaffected.

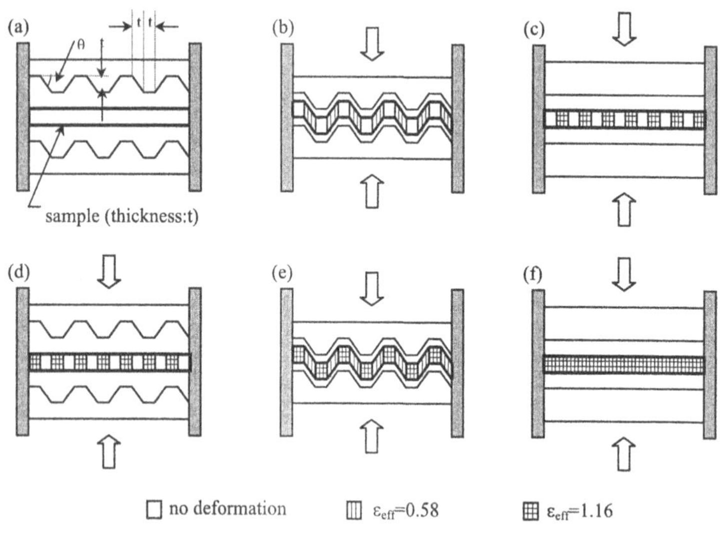

FIGURE 9.2 Schematic illustration of the sequences of CGP technique.

The effective strain (ε_{eff}) in the deformed region after the second pressing becomes doubled (double hatched area in Figure 9.2c). Then the sample is rotated by 180° about the axis perpendicular to plane of the sample (Figure 9.2d). This allows the deformation of the non-deformed region due to asymmetry of the grooved die. In the third pressing, the flattened sample is corrugated again with the help of asymmetrically grooved dies (Figure 9.2e). This induces an effective strain (ε_{eff}) in the previously undeformed region. In the fourth pressing, the corrugated sample is flattened using flat dies (Figure 9.2f). These successive pressings with a pair of grooved dies (Figure 9.2e) and a pair of flat dies (Figure 9.2f) result in a homogeneous effective strain throughout the sample. These four pressings are considered as one pass. Hence, one pass induces a homogeneous effective strain throughout the sample. By repeating this process, very large amount of plastic strain can be accumulated in the sample without changing its initial dimensions and an ultrafine-grained structure can be obtained.

9.3 SHEAR STRAIN AND EQUIVALENT STRAIN

Assuming the following parameters:

Groove angle=θ, Groove width=t, Groove height=t, Interface distance between grooves=t and thickness of sheet sample=t.

The required condition for CGP/RCS is that the groove width and the interface distance between the grooves should be equal. The schematic of shear straining of the sheet specimen is shown in Figure 9.3.

In a single pressing, the specimen element in the shear zone of width 't' gets sheared by a distance of Δx, through an angle of θ. This induces an engineering shear strain, γ_{xy}.

$$\gamma_{xy} = \Delta x/t = \tan\theta$$

If θ=45°, then shear strain is equal to 1. This engineering shear strain of value 1 accounts for an effective strain of value 0.58, as shown in the following equation.

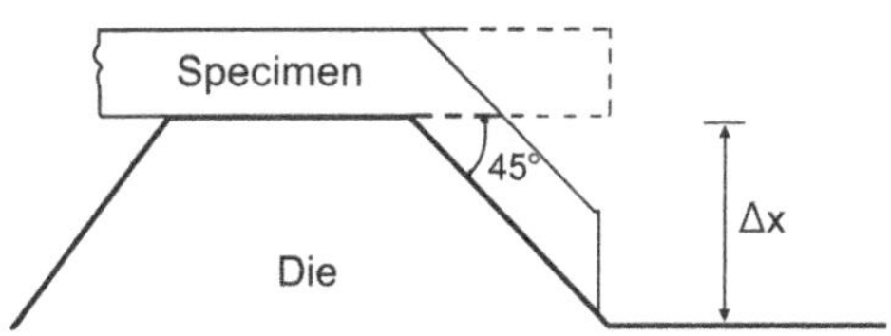

FIGURE 9.3 Schematic of shear straining of sheet specimen.

$$\varepsilon_{\text{eff}} = \sqrt{\frac{2}{9}\left[\left(\varepsilon_x - \varepsilon_y\right)^2 + \left(\varepsilon_y - \varepsilon_z\right)^2 + \left(\varepsilon_z - \varepsilon_x\right)^2\right] + \frac{4}{3}\left[\varepsilon_{xy}^2 + \varepsilon_{yz}^2 + \varepsilon_{zx}^2\right]}$$

Using shear strain $\varepsilon_{xy}=\gamma_{xy}/2$ and assuming the deformation as simple shear with no longitudinal and transverse expansion (i.e $\varepsilon_x=\varepsilon_y=\varepsilon_z=\varepsilon_{xy}=\varepsilon_{zx}=0$), the equation can be simplified as

$$\varepsilon_{\text{eff}} = \sqrt{1/3} = 0.577 = 0.58$$

Hence, it is deduced that a single pressing induces an effective strain of value 0.58 when θ is 45°. Flattening the sheet specimen using a pair of flat dies in the second pressing causes reverse shear deformation in the inclined regions and increases the effective strain to a value of 1.16. In the same way, the third and fourth pressings induce an effective strain of 1.16 in the not deformed regions of the sheet specimen. Therefore, one pass, which consists of four pressings, induces a homogeneous effective strain of 1.16 throughout the sheet sample. Similarly, two passes will induce an effective strain of 2.32; three passes will induce an effective strain of 3.48 and so on. For theoretical analysis, the effective strain is approximated to be homogenous and non-reversible for the sheet specimen, though there is spatial heterogeneity in the actual process, which cannot be avoided.

9.4 MODIFICATIONS OF GP

There are several modifications proposed and executed in the GP. Some of the modified processes are semi-constrained groove pressing, rubber pad-constrained groove pressing, constrained groove pressing-cross route and covered sheet casing-constrained groove pressing.

In semi-constrained groove pressing (SCGP), the sheet specimen is placed between the dies that are tightly constrained by a channel rather than a container as used in the CGP process. In this condition, the sheet specimen gets constrained between the left and right walls, i.e. it cannot move in the longitudinal direction while it is free to move in the transverse direction. The normal, longitudinal and transverse directions are indicated in Figure 9.4.

Another type of GP is Rubber Pad-Constrained Groove Pressing (RP-CGP). Unlike in other GP processes (where two grooved dies and two flat dies are used), RP-CGP process uses only one grooved die and one flat die. The upper grooved die and the upper flat die are both replaced with a flexible rubber pad and a punch, with the rubber pad

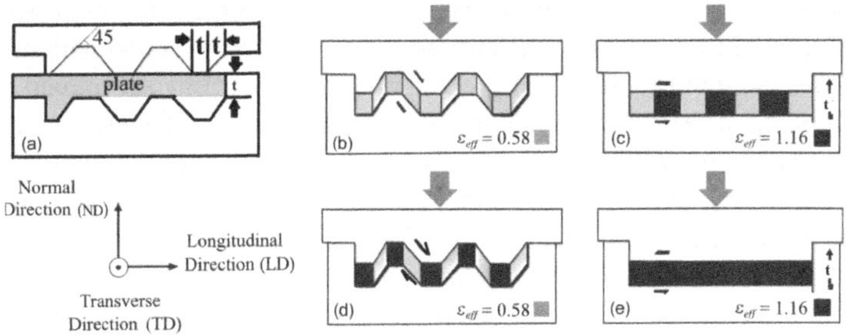

FIGURE 9.4 Schematic of SCGP technique.

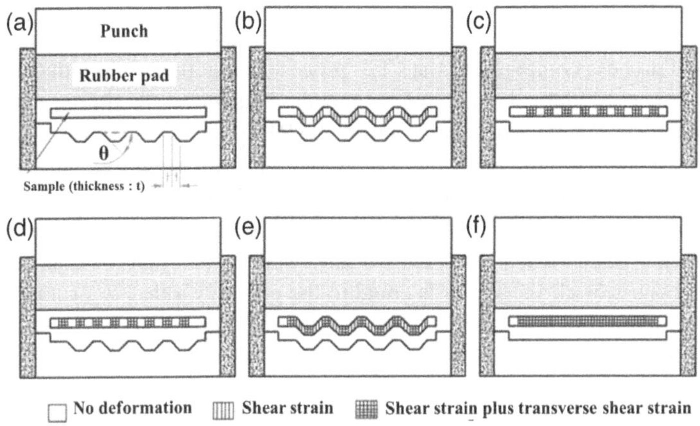

FIGURE 9.5 Schematic of RP-CGP process.

towards the sheet specimen. This is clearly illustrated in Figure 9.5. The rubber pad serves as grooved die for corrugating stages (Figure 9.5b and e) and also acts as a flat die for flattening stages (Figure 9.5c and f). The major advantage is the elimination of one set of die and rubber pad which induces the hydrostatic condition eliminating cracking of the material. Hence the sheet material can be subjected to more number of passes. It is much similar to rubber pad forming in conventional metal forming process.

Other modified types of CGP are constrained groove pressing-cross route (CGP-CR) method similar to the conventional CGP process, two sets of corrugating and straightening dies are used in CGP-CR process as well. The main constraint is that the sheet sample in this process should have a square cross section. Each pass of CGP-CR process includes eight stages – four corrugating and four flattening ones.

Firstly, the sheet specimen is placed such that its rolling direction (RD) is parallel to the die length. The sheet is groove pressed and flattened. After the first two stages of pressing, the sheet specimen is rotated by an angle of 180° around the axis perpendicular to its plane so that the RD is again parallel to die length but in the opposite direction. The sheet is groove-pressed and flattened again. So far, the process is similar to the CGP process. In the CGP-CR process, the sheet is further rotated by an angle of 90° around axis perpendicular to its plane such that the RD is now perpendicular to the die length. Then the sheet is groove pressed and flattened. After these stages, the sheet is rotated by 180° such that the RD is still perpendicular to the die length but in the opposite direction. Again, the sheet is groove pressed and flattened. This whole process is equivalent to one CGP-CR pass, which induces an effective strain double that of CGP. The major benefit of this modified process compared to the CGP process is that the strain induced in the sheet is relatively more homogeneous. This process is similar to that of Cross accumulative roll-bonding (CARB) process.

Another modification of CGP is the constrained groove pressing-covered sheet casing (CGP-CSC) method. In the CGP-CSC method, the thickness of the sheet specimen is less than the groove width. The specimen is sandwiched between two other sheets such that the combined thickness of all the sheets is equal to the groove width. For example, a die with groove width of 1.8 mm is used for pressing a sheet of thickness 1 mm. The 1 mm thick sheet is placed between two sheets of hard material sheet, each with 0.4 mm thickness. The advantage with this method is that it prevents micro-cracking on the surface of the sheet specimens. Micro-cracks develop on the surface of the sheet, which is in contact with the sharp grooves of the die. In this case, the sheets of hard materials are in contact with the sharp grooves preventing the formation of micro-cracks on the sheet's surface. Hence, this method ensures good surface quality for the specimen when compared to the conventional CGP process. This is very similar to hard plate accumulative roll-bonding (ARB (HPR)) process.

Cryo-Groove Pressing (CGP) is another type performed either adiabatically or isothermally. In adiabatic CGP, the samples are immersed in liquid nitrogen during pressing. The samples, including the dies, are immersed in a liquid nitrogen tank for at least 60 seconds before starting the deformation. It is the general expectation that cryo-pressing is more effective in enhancing both tensile strength and yield strength. Contrary to expectations, no significant differences in the microstructures are observed. This may be due to the temperature loss to the environment, which may be high and cannot suppress the dynamic softening

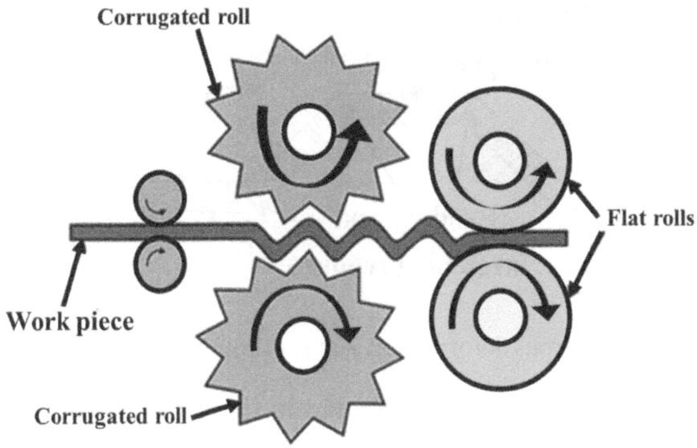

FIGURE 9.6 Continuous RCS process using two sets of corrugated rolls and flat rolls.

mechanisms and the heat due to shear deformation and friction may quicken the loss in temperature. In isothermal CGP, the samples and the dies are completely inside the liquid nitrogen throughout until the entire process is completed. It avoids loss of temperature to surroundings and dissipates the heat produced due to shear deformation. There are not many investigations reported in the isothermal CGP about the microstructure and properties achieved. This may be due to difficulty in the design of dies and process requirements at cryogenic temperature.

Continuous GP machine is capable of producing bulk nanostructured materials. The sample dimension (length) is not limited by the length of the die. A basic GP cycle consists of two steps: corrugation and straightening/rolling. The first step is to corrugate the workpiece, which is performed using a continuous GP facility. The second step is to straighten the workpiece by rolling in a conventional rolling mill, as shown in Figure 9.6.

9.5 EXPERIMENTAL PARAMETERS INFLUENCING THE PROCESSING OF MATERIALS BY GP

9.5.1 Groove Angle

Groove angle (θ), as shown in Figure 9.3, is also an important processing parameter in CGP. Smaller groove angle allowed a greater number of passes. The loading is severe when the groove angle is more and results in the formation of cracks at the corner of the groove, where the material is subjected to the most severe loading condition. Figure 9.7

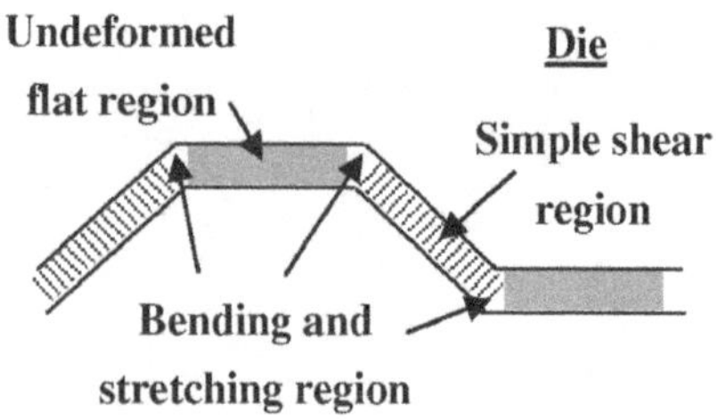

FIGURE 9.7 Different deformation zones of the workpiece.

shows the types of stresses distributed in the sample. Corners are represented as 'bending and stretching region'.

Die with a larger groove angle results in a higher value of hardness. This is due to a greater amount of imposed shear strain in the sheet and consequently, better grain refinement. This greater amount of strain is imposed in the case of die with larger groove angle since shear strength $\gamma_{xy} = \tan\theta$. Shear stress will increase with an increase in groove angle. Hence, ε_{eff} will be greater in the case of die with higher groove angle.

In large angle grooved dies, the mean value of the imposed plastic strain is higher than that of sample processed in lower angle grooved dies, but the uniformity of strain distribution in large groove is less when compared with lower groove angle. Hence, it shows that larger groove angle leads to higher strain values but lesser uniformity in strain distribution. The inhomogeneity factor increases with an increase in groove angle.

9.5.2 Groove Width

Groove width is another important experimental parameter that influences mechanical properties during CGP. Die with smaller groove width allowed less number of pressing cycles as compared to larger groove width. This is because groove width has significant effect on cracking tendency of the sheet. In the sheet specimen, crack always initiates at the place where it touches the corner of the groove (Figure 9.7). At this region, the specimen is subjected to the most severe loading condition. The stress in the specimen at this region is inversely proportional to the groove width and increasing the groove width leads to reduced stress. This decreases the cracking tendency and enables the sheet specimen to undergo a greater number of passes. Similarly in the case of dies with larger groove width allow a greater number of passes for most of metallic sheets. Though the die with larger groove width allows higher number of pressing cycles, the rate of grain refinement is higher for the die with smaller groove width at lower strains. Usually in CGP, groove width will be equal to the thickness of sheet specimen.

9.5.3 Pressing Orientation

Pressing orientation plays a major role during the processing of materials by CGP. There are two modes of pressing procedures – single-orientation pressing and cross-orientation pressing. In single-orientation pressing, the sheet is rotated by 180° about the z-axis after every odd number of pressing cycles. In cross-orientation pressing, the sheet is rotated by 180° about the z-axis after every odd number of pressing cycles and in addition, the sheet is rotated by 90° about the z-axis after every even number of pressing cycles. Single-orientation pressing is the same as that of the conventional CGP procedure, and cross-orientation pressing is the same as the CGP-CR method.

Cross-orientation pressing procedure has allowed more number of pressing cycles without crack formation and leads to fine grains as compared to single-orientation pressing due to the more even distribution of the bending and stretching locations, where the material is subjected to the most severe loading condition. Therefore, cross-orientation pressing led to higher accumulated strains and finer grain sizes. By rotating the sheet by 90° about the z-axis after every two consecutive pressing cycles, the cross-orientation pressing procedure alters the maximum shear-stress plane back and forth from being parallel with the x–z plane to being parallel with the y–z plane. This introduces the complexity of the deformation field, which enhances the rate of grain refinement. However, the grain size resulting from single-orientation pressing approaches the grain size obtained via cross-orientation pressing at relatively large strains. Similar phenomenon was observed for the dies with larger and smaller groove widths. The steady-state grain sizes attained at higher strains is due to dynamic balance between dislocation generation and recovery. The rate of dislocation recovery is mainly controlled by the deformation temperature and strain rate, and both pressing orientations are expected to have similar recovery rates. Therefore, similar grain sizes from both the pressing orientations at large strains suggest that the rates of dislocation generation are similar for both of them. This means that single-orientation pressing, although having a lower rate in storing dislocations at lower strains, can accumulate sufficient dislocations with the increase in strain. Therefore, single-orientation pressing catches up with cross-orientation pressing in the dislocation generation rate and the same grain size is exhibited by both the procedures at large strains.

The die with a larger groove width in combination with the cross-orientation pressing procedure yields the best results. Also, with larger groove width in combination with a cross-orientation pressing procedure led to a very uniform hardness profile, which is even higher compared with the uniformity of the pre-deformed specimen.

9.5.4 Friction

Surface condition of sheets during deformation, i.e., friction between the dies and the sheet is an important parameter in micro-cracking mechanism. Use of Teflon layers as a lubricant promotes the surface condition and delays micro-cracking and allows for more number of passes. Among the four conditions, namely: no lubricant (l=0.25), machine oil as lubricant (l=0.15), molybdenum disulphide as lubricant (l=0.12–0.15) and Teflon layers as lubricant (l=0.04–0.12), where 'l' is the friction coefficient. Teflon layers yield the minimum friction between the dies and the sheet allowing more passes of CGP. Homogeneity of microstructure is the maximum in the case of Teflon layers. The uniformity of strain distribution decreases with the increase in friction coefficient.

9.6 ADVANTAGES AND LIMITATIONS

GP technique has an advantage over other SPD techniques, such as ECAP and HPT, due to its applicability to sheet metals. ARB is an effective technique for processing plate-shaped materials, but it has a disadvantage not encountered in GP, namely it involves a repeated bonding between two plates. ARB has strict requirements for surface condition of rolled plates and the atmosphere during the roll-bonding process. If a perfect bonding is not achieved, then the mechanical properties of the materials degrade due to the presence of bonding interface. Comparing the GP and ARB processes based on experimental results, the final grain size of alloy obtained after three passes of GP is smaller than the grain size for the same alloy subjected to eight passes of ARB.

Though GP is a good technique, it has some limitations. The sheet metals can be subjected to only a few passes. After certain number of passes, due to the sharp profile of the dies, micro-cracks appear on the surface of the specimen, and they lead to decrease in strength and other mechanical properties. GP not only decreases the ductility but also the toughness of the materials. The grain refinement rates are lower in the

GP technique when compared to other SPD techniques. This is due to the lack of multiple shear-stress planes during deformation and partial annihilation of dislocations due to reversal of loading direction during GP. The drop in strength is observed at lower strains during GP when compared with other SPD processes such as ECAP and HPT. This phenomenon may be due to the difference in the deformation paths and the larger free surface to volume ratio for sheet materials when compared to bulk materials processed by ECAP and HPT. Free surfaces act as effective dislocation annihilators and hence higher dislocation annihilation is observed at lower strains during GP. Fatigue cracks developed on the surface of the workpiece limit the number of GP cycles that can be applied to a workpiece without failure.

10 Friction Stir Processing

10.1 PRINCIPLE AND PROCESS

The basic principle of friction stir processing (FSP) is to allow the base metal to undergo severe plastic deformation (SPD) through the strong stirring action of the rotating tool to improve the densification, homogenisation and refinement of the microstructure. The refinement and homogenisation of grains during the process is the result of intense local plastic deformation and stirring heat caused by FSP. The improvement in microstructure and mechanical properties eliminates the defects of the substrate.

FSP utilises a special non-consumable rotating tool having a cylindrical cross section with a shoulder and pin. The rotating tool is plunged into the metallic surface and travels along the surface of the material at a pre-defined speed as shown in Figure 10.1a. The shoulder applies pressure on the material, while the pin creates a rotating vortex that causes SPD called as the stir zone (SZ) due to friction resulting in significant microstructural changes. The required thermal energy to soften the material is produced by the friction between the metal surface and the rotating plunged shoulder. The tool pin length controls the depth of the processing regime. Therefore, it can be regarded as a material processing technique for severe plastic deformation (SPD). During FSP, heat is generated due to the friction between the top of the work piece and the tool shoulder as the pin descends on the work piece and the shoulder contacts the work surface. Consequently, plastic flow occurs due to material softening from heat generation, leading to different zones with inherent microstructural characteristics. These zones are divided into four regions: SZ, thermo mechanically affected zone (TMAZ), heat-affected zone (HAZ), and base material (BM) as shown in Figure 10.1b.

Due to SPD and frictional heat occurring simultaneously, the SZ of the specimens made by FSP displays an equiaxed and fine microstructure. The FSP process leads to significant plastic deformation of the material resulting in the grains breaking down during plastic deformation, creating a lot of low angle, misoriented grain boundaries

DOI: 10.1201/9781003596110-10

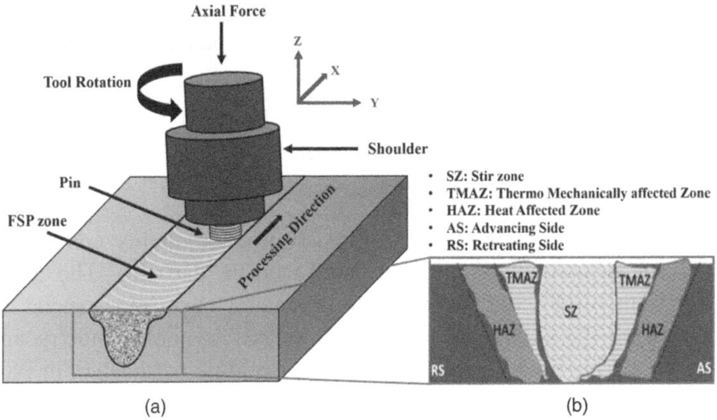

FIGURE 10.1 Schematic illustration of the (a) FSP technique and (b) the different macrostructural zones created during FSP.

and extremely favourable nucleation locations during recrystallisation. There appear to be two main causes for variations in grain size: (i) heat input that results in annealing, leading to grain growth; (ii) continuous dynamic recrystallisation brought on by the tool pin's stirring action, producing new nucleation sites and reducing the grain size. During dynamic recrystallisation (DRX), low-angle boundaries become high-angle ones, and new grains form at preferred sites. Once the fine-grain nuclei have begun to form, a microstructure with fine, equiaxed grains is produced. The processed regime exhibits enhanced hardness, strength and wear resistance. Indeed, this process is successfully used to alter the surface properties of non-ferrous alloys such as aluminium, magnesium and copper.

10.2 PRODUCTION OF COMPOSITES BY FSP

Adding reinforcements during the process enables FSP to produce composite materials. By controlling different parameters of FSP one can produce bulk, surface and FGM composites. FSP-produced composites overcome the drawbacks of liquid and powder metallurgy routes. The inherent characteristics of the process help in homogeneous mixing, better consolidation, and grain refinement. Further, the controlled heating assists in the formation of desirable intermetallic phases. Some common examples of MMCs fabricated via FSP are Aluminium metal matrix composites (AMMCs), Magnesium metal matrix composites, Copper metal matrix composites, Steel metal matrix composites, etc.

The method employed to insert the secondary phases into the base metal is a critical factor affecting secondary phase distribution and consequently the matrix properties in FSP.

10.2.1 Methods for Incorporating Reinforcements in Composites

A good reinforcement strategy helps to produce quality composites with the desired distribution of reinforcement particles. The insertion of the reinforcing particles into the matrix material is crucial for fabricating the composite material. Besides this, tool geometry and processing parameters are considered as deciding parameters in FSP. Reinforcement volume in the SZ predominantly determines the final properties of the composite. Different reinforcement deposition methods are used in FSP. Every reinforcement method has its own merits and demerits regarding reinforcement distribution, microstructure, thickness and quality of fabricated composite.

10.2.1.1 Groove Filling Method

In groove filling (Figure 10.2a) method, narrow single or multiple grooves in the base metal sheet or surface are cut and filled with reinforcements and FSP is utilised to develop the surface composite. During the processing of grooved specimen, the reinforcement particles may come out resulting in improper fabrication of the composite. To overcome this problem, FSP tool without pin (Figure 10.3) is used to cover the groove and enclose the reinforcement, and this is known as closed method. The friction between the pin-less tool and the base metal surface leads to heat generation, and the surface of the groove is closed to prevent the secondary particles from flying out during the FSP process.

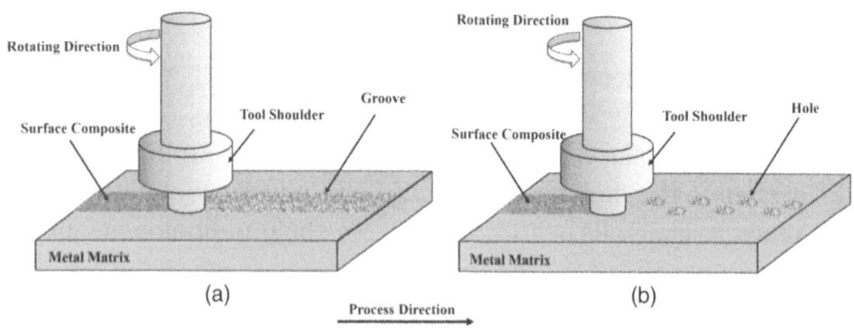

FIGURE 10.2 Schematic illustrations of the FSP process by (a) groove filling method and (b) hole filling method.

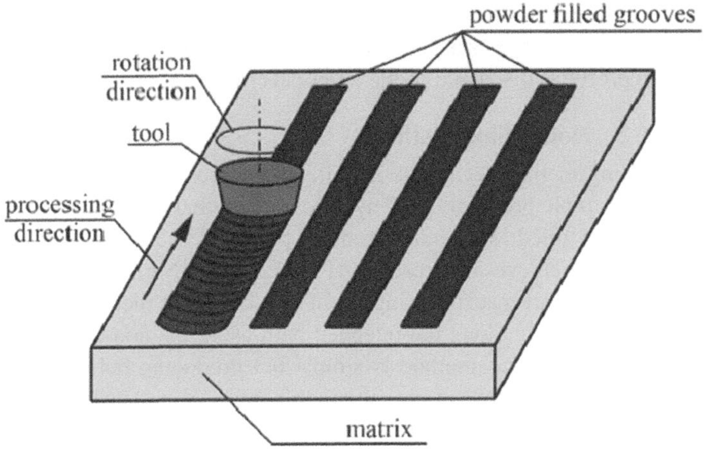

FIGURE 10.3 Graphical representation of groove method with pin-less tool.

Metal adhesive tape to cover the groove is also used. Processing of the covered groove specimen produces composite with minimal defects. The composite plates made using the multi-groove approach received more heat than the ones made using the single-groove technique; hence, they had higher heat transfer through the plate depth. Higher heat transfer could contribute to a high-temperature gradient through the plate, allowing suitable mixing of plasticised base metal in vertical and rotary routes, resulting in a smooth crown. Additionally, it can be deduced that the heat generation increases incrementally by raising the number of grooves.

Mostly, grooves on the surface of the base metal are made as straight lines along the traverse direction. However, recently grooves perpendicular to the traverse direction are also used to avoid exclusion of the coating material and to better combine the coating material with the matrix.

Generally, the groove particles width (w) should be less than the pin diameter (d) and the most recommended ratio width to pin diameter (w/d) is 0.5 to fabricate a defect-free surface composite along with better dispersion of reinforcement particles. Beyond this limit, there is a high tendency of defect formation and nonuniform distribution of reinforcement particles. And also, the distribution of reinforcement particles on the surface became more uniform with the increase of groove depth. The SZ is reduced with increasing groove width. It results in the composite area reduction due to less plasticised material and increases the reinforcing particles. The square pin profile offers more uniform distribution of particles as compared to other pin profiles. The effect of the pin profile is discussed in the latter part of the section.

10.2.1.2 Hole Filling Method

Hole filling method technique is quite simple but requires significant time for sample preparation. Tiny holes are drilled on the base metal surface and filled with reinforcement particles usually mixed with methanol to remove air presence (Figure 10.2b). Similar to the slot method, holes are packed using a pin-less tool, and then FSP is performed lengthwise along the intended path using the non-consumable tool with a pin. This method is simple but producing holes over the base metal is time-consuming, although compared to the slot method, more powder can be added, and escaping of the powder is reduced. This method is more suitable when more powder is required during processing.

The parameters such as hole dimensions and pattern of holes (like rectangular grid, zigzag and offset holes) as well as the spacing between two holes, enable control of the distribution of reinforcement particles in the base plate. The distribution of reinforcement particles is more homogeneous in case of zig-zag blank holes. Blank hole diameter is generally taken as less than half of tool pin diameter.

Blank hole FSP reduces the possibility of uncontrolled movement of the reinforcing phase beyond the modified area. This increases the possibility of uniform distribution of reinforcement particles in the process zone. Similar to the groove filling method, it can be used with a pin-less FSP tool to close the holes with the use of the generated heat, and then, with a pin tool, carry out FSP to fabricate the surface composite.

10.2.1.3 Sandwich Method

In the sandwich reinforcement strategy, reinforcement particles are placed in between two plates, making a layer structure. Then, FSP is performed on the specific path. During the processing of the plasticised base metal, the particles are mixed and make a composite layer. This is a simple strategy allowing fabrication of bulk composites/MMCs. Sometimes this sandwich combination of two base plates and reinforced particles is covered with tape along sides of plates to seal the reinforced particles. This reinforcement strategy has the capacity to increase the amount of reinforcement phase in composites by increasing the number of layers. The pre-processing time in this strategy is very less in comparison to hole and groove strategies for the fabrication of bulk composites. The high thermal energy created by the tool is responsible

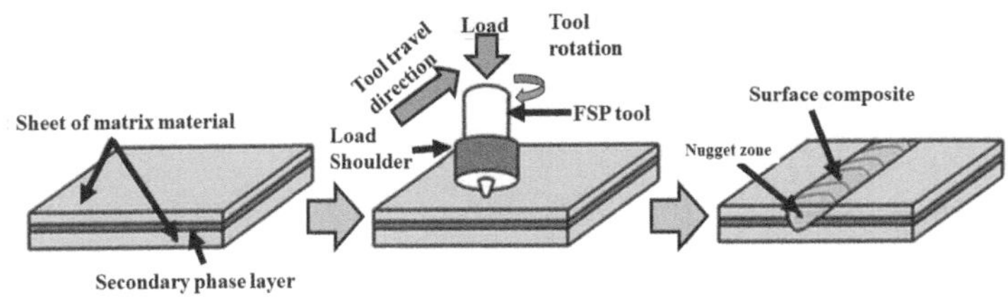

FIGURE 10.4 Sandwich method procedure.

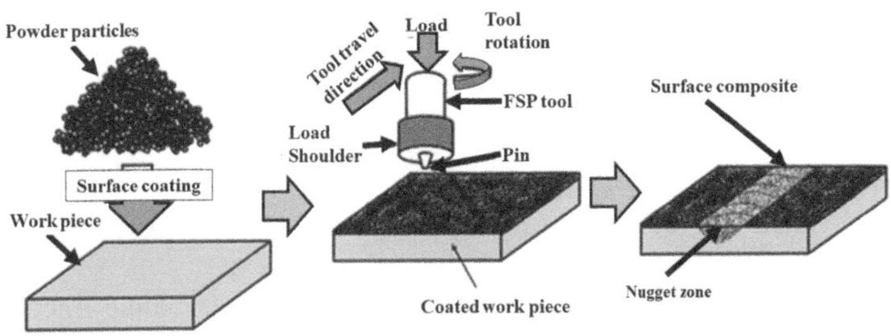

FIGURE 10.5 Surface coating method procedure.

for breaking the reinforcing particles, which helps the manufacture of composites. Figure 10.4 shows the procedure of the sandwich method.

One challenge associated with this process is the placement of the reinforcing particles. For the uniform dispersion of the particles, the multi-pass processing is needed. The particle size selection depends on the FSP process parameters and tool design.

10.2.1.4 Surface Coating Method

In this method, pre-mixed particle paste is applied on the base metal surface by the coating method. Then, FSP is performed on the specific path. Similar to the sandwich process, this process also does not require any capping pass and machining. One advantage of this process is that the whole material can be converted into a composite using multiple passes. Figure 10.5 shows the procedure of the surface coating method.

Air Plasma Spray (APS) technique or High Velocity Oxy Fuel (HVOF) techniques are also used to coat the particles on the surface. After the powder is coated on the substrate, the plates are subjected to FSP. The distribution of reinforcement particles can be controlled by APS and HVOF more effectively than by other methods. This strategy is found to be more suitable for nano-size reinforcement particles as compared to other available reinforcement strategies of FSP with

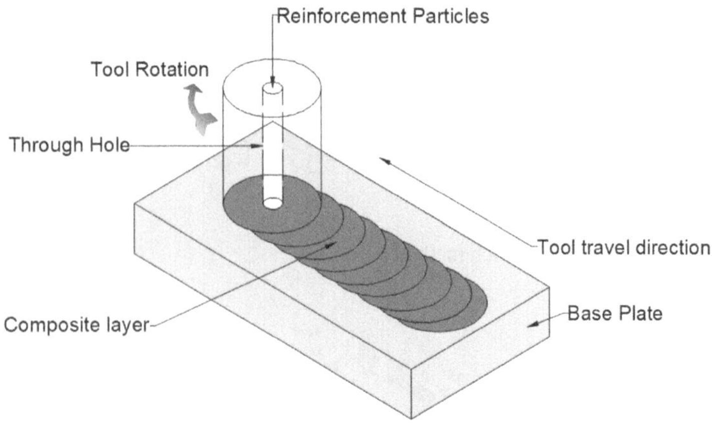

FIGURE 10.6 Schematic representation of DFSP.

excellent properties such as homogeneity, control of volume fraction, and fine-grain microstructure.

10.2.1.5 Direct Friction Stir Processing

Direct friction stir processing (DFSP) involves a hollow non-consumable rotating tool with shoulder as well as with or without a pin filled with reinforcement particles. During processing, the reinforcement particles flow out from the hole in the tool and enter between the shoulder and the base metal. Frictional heat and relative motion of the tool help to mix the reinforcement particles in the matrix and form the surface composite. The solid-state nature and the mechanical stirring of the tool help to achieve homogenisation of powder/particles within the substrate. To retain reinforcement particles within the hollow tool before processing, the opening is generally sealed with a foil. The process is associated with one challenge, which is to make the hole into the tool. A precise hole is needed to insert the reinforced particles and still maintain enough strength for the processing. Schematic representation of DFSP is shown in Figure 10.6.

DFSP is more suitable for producing surface composites. DFSP also refines the grain size significantly resulting in an improvement in surface hardness. In comparison to blank holes and grooves, DFSP is less time-consuming because it does not require surface preparation before processing. DFSP is suitable to produce thin layer of surface composite and can control the composition of reinforcement particles during the process. DFSP lags in producing bulk composites and bulk FGM;

however, it can produce surface FGM by varying the composition of the reinforcement particles during processing.

10.3 MODIFICATIONS IN CONVENTIONAL FSP

Generally, the heat input into the processing regime increases with the rotational speed of the tool and decreases with tool travel speed. By controlling these parameters, the heat input can be controlled to obtain the desired microstructure. Besides controlling heat input, a proper cooling rate is very helpful to obtain an ultrafine-grained microstructure.

10.3.1 Water-Cooled FSP

To enhance the grain refinement effect during FSP, an effective method is to perform the entire process in a liquid cooling medium, such as flowing water or liquid nitrogen. High-efficiency cooling will help fast heat dissipation during FSP avoiding grain growth due to frictional heating. Water-cooled FSP is schematically shown in Figure 10.7. This process is also called Underwater FSP or Submerged FSP. To transfer the heat generated between the tool and the sample during the FSP process as quickly as possible, a copper mould is normally used as a backing plate. On the one hand, a water tank is arranged on the upper surface of the copper mould, which can ensure the flow of cooling water. During FSP, flowing water at room temperature always floods the work piece surface. The rate of flowing water can be regulated using the valves in the water inlet and water outlet.

The results show that water-cooled FSP (WFSP) has unique advantages in improving the microstructure uniformity of materials and also

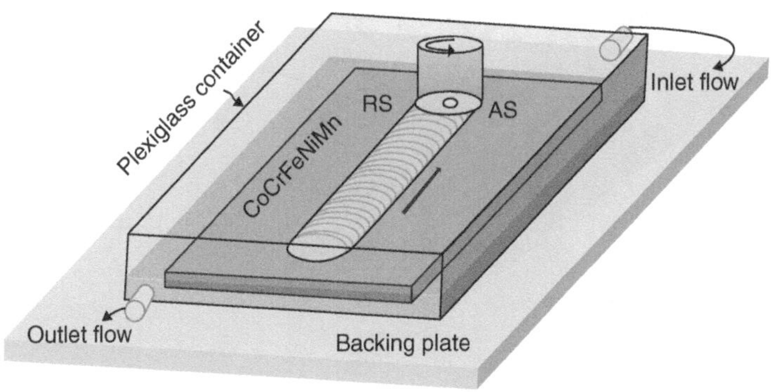

FIGURE 10.7 Schematic of water-cooled FSP setup.

providing a more symmetrical cross-section at the macroscopic level as compared to conventional FSP. However, the microstructures of the surface layer are more uniform and refined as compared to the inner layer as the surface is in direct contact with the flowing cooling water and cools rapidly.

10.3.2 Cryogenic Friction Stir Processing

The other way for enhancing the fast heat dissipation is to use FSP at lower temperatures. Cryogenic friction stir processing (CFSP) technique involves submerged liquid nitrogen cooling and a non-shoulder stir tool. A high fraction (<40%) of nanograins can be obtained during CFSP, resulting in a synergistic enhancement of tensile strength and ductility.

10.3.3 Friction Stir Vibration Processing

Additional heat energy during FSP to help soften the material in the vicinity of the tool to decrease the processing loads (tool torque and axial load), enhances the material flow, dispersion of the added particles, modifies grain size, and improves joint efficiency. As mechanical energy, ultrasonic vibration has been regarded as special for its remarkable ability to soften the materials under plastic deformation without appreciably melting the joints used in friction stir vibration processing (FSVP). Schematic diagram of FSVP is shown in Figure 10.8. This

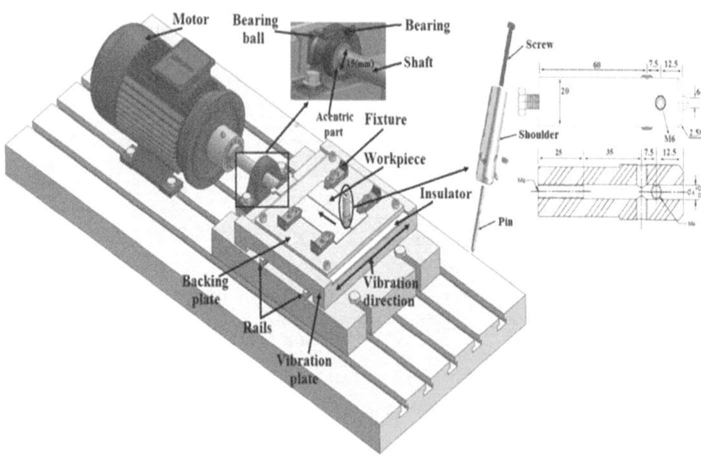

FIGURE 10.8 FSVP process set up.

FSP setup is fixed on the vibrating plate, and the specimen is vibrated normal to the FSP line as the motor is turned on. Vibration frequency is controlled by a driver attached to the motor. Normally vibration frequency is 50 Hz with an amplitude of 0.5 mm.

Strength, ductility and wear resistance of FSVP specimens are higher due to smaller grains and more homogenous distribution of the strengthening particles as vibration is applied. The top surface of FSVP is smoother due to less flash produced as compared to conventional FSP due to the improved material flow because of higher straining of materials. The flow-related defects including voids and tunnel are absent due to the application of vibration during FSVP. In the fabrication of surface composites, the dispersion of the reinforcements is better with no clusters due to sufficient material flow and enlarged SZ. Additionally, the tool torque is reduced nearly 10%, and the material flow is more stable and uniform in the SZ.

10.4 EFFECT OF PROCESS PARAMETERS

FSP is a complex process involving heating, plastic deformation and metallurgical interaction. Thus, it is necessary to identify the effect of each parameter to achieve the desired surface enhancement. These parameters can be classified as machine parameters, tool parameters, passes and reinforcement strategies. These parameters can affect the heat input and the resultant properties.

10.4.1 Machine Variables

During FSP, machine variables are responsible for the quality of the MMCs. Among the Machine parameters, the rotational speed (RS), the traverse speed (TS) and the number of FSP passes are the most important parameters that affect the microstructure and mechanical properties. Usually, higher heat input results in grain coarsening, while lower heat input may lead to metal defects such as porosity, cracks, etc. TS is important for the distribution of reinforcement particles during composite preparation. Excessive TS results in inadequate plastic flow leading to nonuniform distribution of the reinforcement particles. The heat generated by lower RS but higher TS could contribute to grain refinement and metal strengthening. With higher RS but lower TS, less pin travel per revolution results and thus produces excessive heat, resulting in possible hardness deterioration and grain coarsening.

The friction stir process with one pass refers to FSP in which the tool goes over the processed material only once. In a single pass processing, the reinforcement particles have a greater chance of aggregating into the processed zone leading to poor surface quality in the composite. More than one pass is required to increase the homogeneity of the reinforcing in the BM and to eliminate the joint defects. The multi-pass FSP can be applied by performing FSP on the same seam in the same direction, the same seam in different directions, or in an overlapping manner with 50% overlap on the previous original seam. As the number of FSP passes rises, there is an observed increase in the fragmentation of particles. This results in a progressive reduction in the average size of particles and enhances the inhibitory effect of particles on grain boundary migration during recrystallisation. However, it is crucial to consider that excessive fragmentation of particles can weaken load transfer strengthening, thereby diminishing the overall strengthening efficiency of the composite. As a result, a significant effect can be had on the quality of the final processed matrix in terms of tensile strength, microstructure, microhardness and wear resistance. Figure 10.9 illustrates the impact of RS and TS on the occurrence of defects in FSP.

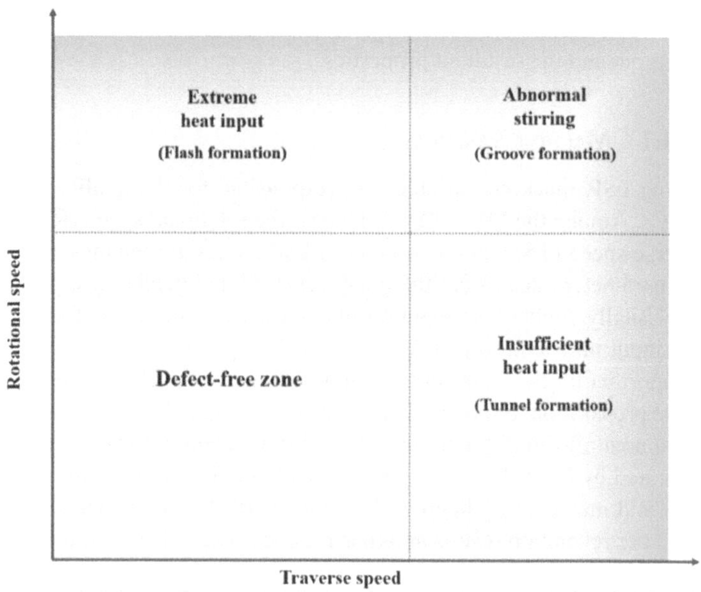

FIGURE 10.9 Influence of FSP parameters on the defects.

10.4.2 Tool Pin Profile

In FSP, the tool is essential for altering material flow and the resultant microstructure, among other process parameters. In FSP, the tool geometry - inclusive of the probe shape and size, shoulder configuration, and diameter- is an important parameter that controls heat generation, material flow, matrix and particle blending, microstructure, and characteristics of the SZ. The principal source of heat during tool plunging is friction between the pin and the work piece. Most of the heat is generated when the shoulder contacts the work piece surface because of friction between them while additional heating is caused by material deformation. The different pin types employed for FSP are depicted in Figure 10.10.

In the case of metals and alloys, square and triangular pin profiles produce finer microstructure and higher mechanical properties. Also, no material will be collected at the tool outline. The flow of material happens at the progressing side to the withdrawing side and create miniature worm shape. Since square and triangular tool profiles have sharper corners, they induce effective processing of the materials similar to that of ECAP. Triangular tool pin results in improved mechanical and wear behaviours over a square tool profile. A tapered tool pin with screw threading offers the best mechanical and microstructural qualities. The tool pin thread pitch plays a significant role in producing a process zone devoid of defects. A stepped, conical, threaded pin enhances material flow, enhancing mechanical characteristics, while effective material flow is provided by a polygonal pin profile.

In case of composites, compared to the square profile, straight cylindrical profile improves the surface quality, unifies the dispersion of the reinforced particles, and lowers the heat input because of the smooth

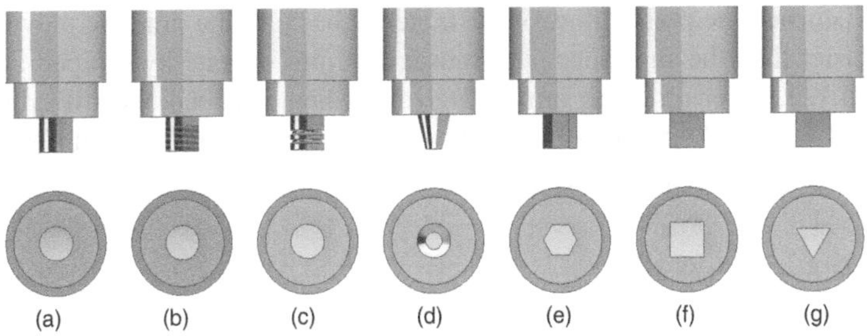

FIGURE 10.10 Pin shapes for FSP tool: (a) straight cylindrical; (b) cylindrical threaded; (c) cylindrical grooved; (d) straight conical or tapered; (e) straight hexagonal; (f) straight square; (g) straight triangular.

geometry, resulting in finer grains. Besides regular pin shapes, hybrid pin having half threaded and half-square shape is also used widely. Samples processed with the hybrid tool show improved strength compared to conventional tool due to finer grains produced by the hybrid tool. Threaded conical pin produces the best surface quality compared to the other pin shapes. Threads around the pin mix the soft material properly producing SZ having more turbulence. The eccentric tool pins provide good flowability and intermixing of particles with matrix, increasing the size of SZ compared to SZ with concentric pins (circular and square), and the grains in the SZ are sufficiently finer with the eccentric tool pin than with the concentric pin. Maximum mechanical properties are achieved with 1 mm eccentric cylindrical tool pin.

Ultimately, defect-free process zones can be produced using pin profiles that are square, conical, triangular straight-cylindrical, tapered-cylindrical and cubic. Despite the effectiveness of the tool pin design, a shoulder-to-pin diameter ratio of three could result in the ideal microstructure.

10.4.3 Multi-Pin Tools

The multi-pin tool has a greater refining impact on the particles even at slower TS because the multi-pin tool is better at stirring and creating frictional heat. Figure 10.11 shows the impact of the single and multiple-pin methods on the plastic flow during the FSP process. In the single-pin method (Figure 10.11a and b), the moving pin splits the plastic flow of material into two pieces, which are then deposited behind the tool. However, in the multiple-pin method (Figure 10.11c–e), the offset pins on the tool cause the movement of each pin and tool to be different. Furthermore, the offset pins divide the plastic material flow into many strands. As can be seen in Figure 10.11d and e, the initial pin disturbed the plastic flow of the material because the pins are moving faster than the substance, and subsequent pins repeatedly scraped and stirred it. For instance, immediately after being rubbed and stirred by pin A, flow A and flow B would be stirred again by pin B, and flow C would be stirred again by pin C as well.

10.4.4 Heat Input

Heat input during FSP affects grain growth and dynamic recrystallisation, which determine the final properties. Hence, precise calculation is necessary. The total heat generation (Q) during FSP in its general form is given below:

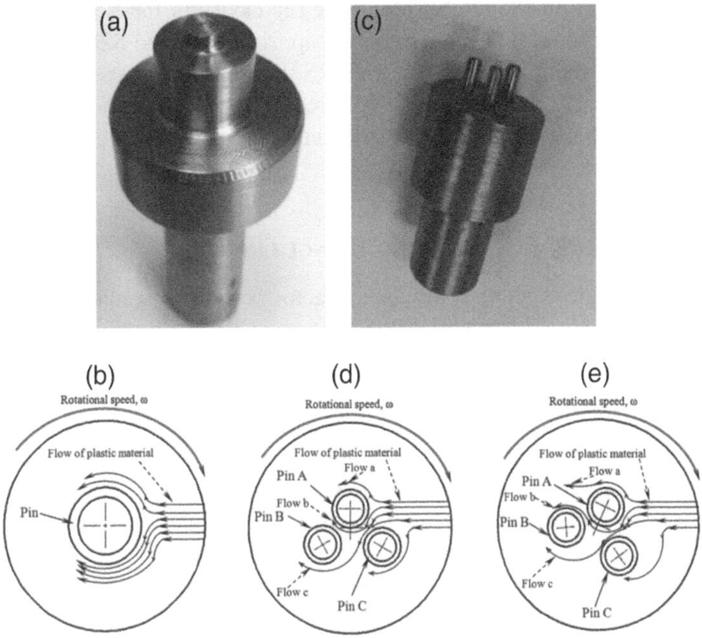

FIGURE 10.11 Images of the FSP tool using a single pin (a), multiple pins (c), and the impact of the pin model on the flow of plastic material (b–e).

$$Q = \frac{2}{3}\pi\omega\mu\tau\left[R^3 + X \times r^2 \times h\right]$$

where ω- *tool rotational speed,* μ *- friction coefficient, τ - shear strength, R -pin diameter, and* *X* is calculated depending on the tool pin profile and is 3 for a straight cylindrical pin profile, r is the pin diameter at the root and h is the pin height. Heat input (*Q*) directly correlates with RS and inversely relates to processing speed. Optimisation of rotational and traverse speeds leads to optimal dispersion of the powder and a reduction in microstructural defects. Heat input rises as the RS grows or the process speed declines, leading to a greater peak temperature. Increasing processing speed means less time exposure to above recrystallisation temperature; thus, with growing rotational and process speeds, the peak temperature, like heat input, rises and falls, respectively. In addition, heat input changes the SZ shape. Generally, lowering the rotational speed is bound to result in a basin-shaped SZ alongside the cavity in the RS zone due to low heat input. On the contrary, trapezoidal-shaped SZs could be obtained without any cavity at high RS. During the process, it should be noted that "tool dragging" might result from low plastic deformation of the base metal and an inadequate amount of plasticised material to fill the space behind the pin.

10.4.5 Tool Tilt Angle and Plunge Depth

Tool tilt angle is another influential factor besides rotational and processing speeds. The material flow pattern in SZ during FSP, effective stirring action, and frictional forces are all enhanced by raising the tool tilt angle from 0 to 2°, resulting in a considerable change in heat generation. Extensive specific surface contact between the tool and the work pieces when the tool tilt angle is 0° results in excessive heat during FSP. The plasticised material is pulled away from the centre of the SZ by excessive heat in the stir region. At greater tilt angles, due to effective forging and stirring action, the plasticised material flow is sufficient. Less porosity will result from effective forging and stirring, so fine, homogeneous grains will be developed.

The tool/work piece contact area is determined by the tool tilt angle and plunge depth. Increasing the tilt angle while maintaining the plunge depth causes less heat generation due to a reduction in the interface area. Increased plunge depth, on the other hand, compensates for the reduced contact area and heat generation. Tool tilt angle improves material flow from the tool's front to backside, resulting in better SZ filling. A small tilt angle causes voids in the processed region, and large angles can also cause voids. In case of composites, by increasing the tool tilt angle from zero to two degrees, due to the effect of the frictional forces and heat generation, which improve the compaction and prevent the accumulation of reinforced particles and improve the distribution of the particles in the SZ, resulting in increase of the hardness of the composite. Large tool tilt angles involve a variety of faults, including tunnelling problems. A large tool tilt angle results in more SZ thinning, which reduces the tensile strength significantly. The optimum angle will be 1°.

As the plunge depth reduces, the strength is also lowered since when the plunge depth is less, the least contact between the shoulder and the work piece, leading to insufficient material mixing, and heat transfer. As the plunge depth increases, the strength increases. However, above a critical limit, the tensile strength decreases, and elongation increases. This is due to the excessive heat generated and concentrated primarily in the SZ.

10.5 ADVANTAGES AND LIMITATIONS

Tribological, corrosion, and erosion properties are most affected by surface properties. Extensive plastic deformation of the surface alters the hardening, grain refinement, and improvement of the formation of passive layers of the surface significantly. FSP is the best method to alter the surface properties. Notably, FSP is not limited by the shape of work piece and the processing environment. The most unique advantage of FSP is its ability to form MMCs on the desired surface with improved bulk and surface properties, which promote the durability and functionality surpassing that of conventional metals and metallic alloys. Besides, fusion welds are also stirred by FSP to replace the cast structure with a fine, worked structure, removing any weld defects, enhancing the properties. Moreover, micro/nanocomposites fabricated by other techniques are also processed by FSP to obtain much stronger bonding with the base metal compared to other modification methods. Powder products of unique composition can be brought to full density with attractive strength and ductility. Thus, FSP has the potential to be applied in many applications such as aerospace, automotive, marine, biomedical and other engineering fields.

The limitations of FSP include: the need for precise clamping and tooling, potential for defects like "worm holes" if parameters are not optimised, limitations in complex geometries, relatively slow processing speeds, high tool wear, and the requirement for specialised operator skills, making it less suitable for large-scale production of intricate parts in certain materials. FSP can be challenging with materials that have high flow stress or poor thermal conductivity. Achieving optimal results requires careful control of processing parameters like tool rotation speed, TS, and clamping force, which can be challenging to maintain consistently. The specialised machinery required for FSP is expensive, impacting the overall cost of production. Skilled operators are needed to accurately control the process and to identify potential issues.

11 Microstructure and Properties of Materials Processed under SPD

11.1 GRAIN REFINEMENT MECHANISM IN SEVERE PLASTIC DEFORMATION

Microstructure refinement is an important phenomenon that takes place during severe plastic deformation (SPD) of the material, where the grain size gets refined, and the properties of the material improve. SPD process refines the grain size of the materials to the sub-micrometre range, and sometimes even to the nanometer range due to a high amount of strain imposed by subjecting to repeated cycling, and the amount of total strain is the geometrical addition of strain imposed in each pass. The grain refinement mechanism depends on the method of SPD and the properties of the materials. Shear localisation and shear band formation depend on various material parameters, especially crystal structure, which determines the availability of slip systems as well as their interactions and in turn dictate strain inhomogeneity in materials discussed in Chapter 1 briefly.

Cubic BCC and FCC metals and alloys show enhanced strain hardening due to the availability of a large number of slip systems, which usually delay the shear banding. However, the presence of microstructural heterogeneities or differential strength between constituent phases on the other hand assist by setting up shear localisations and shear band formation. Due to the limited number of independent slip systems in HCP metals and alloys, slip activation leads to strain inhomogeneity among the grains depending on their orientation with respect to the loading direction. HCP metals are more likely to undergo shear bands. However, slip activation for HCP metals and alloys depends on the c/a ratio. For example, while basal slip activates Mg due to a near ideal c/a ratio (~1.63), prism slip dominates the deformation of Ti at room temperature due to its lower than the ideal c/a ratio of 1.5678. So, metals

DOI: 10.1201/9781003596110-11

with a low c/a ratio have a higher inclination for strain hardening and correspondingly lower tendency towards shear band formation. Shear banding also occurs due to complex interaction between adiabatic heating and resulting thermal softening with slip/twining activation and strain hardening usually after high strains. Shear band formation or strain hardening influence towards shear band formation and the resultant microstructure and texture evolution in SPD. Stacking fault energy is another important property that has definite impact on shear band formation, especially for FCC metals and HCP metals. The grain refinement by shear band formation and strain hardening is discussed briefly in the following sections.

11.1.1 Grain Refinement by Shear Band Formation

During the first pass, the deformation bands developed around 35°–45° of the direction of loading as shown in Figure 11.1. During the second pass, when the sample is rotated lead to formation of new deformation bands with almost the same orientation with respect to the loading direction as developed in the first pass, and these new deformation bands start intersecting with the deformation bands created in the first pass. Again, the third pass led to the formation of new deformation bands with the same orientation with respect to the loading direction, and these new deformation bands severely intersect with all the previously

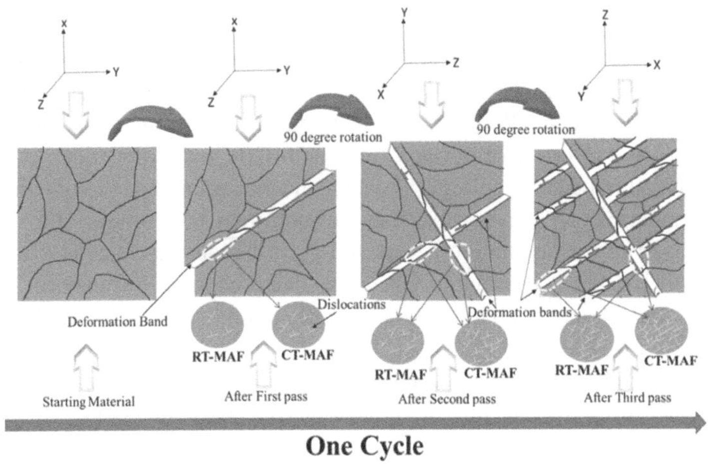

FIGURE 11.1 Grain refinement mechanism.

created deformation bands. Deformation bands crossing each other subdivide the grains into a large number of sub-grains. The boundaries of sub-grains rearrange themselves and convert into high-angle grain boundaries; thus, all sub-grains transform into grains. During the SPD, as strain increases, this leads to the development of dislocations within the deformation bands. These dislocation movements are hindered by the secondary phase particles and by the grain boundaries. The dislocations pile-up at these two places leads to the formation of dynamic recrystallised grains. The severity of deformation bands is influenced by the processing temperature during the SPD. SPD at room temperature or at warm temperature leads to annihilation of dislocations, and dislocation density reduces within the deformation bands, and this further promotes the softening during the processing. In the case of SPD at the liquid nitrogen temperature, it leads to suppression of dislocations annihilation and enhances the dislocation density within the deformation bands. These highly deformed deformation bands are the main reason for severely deforming the structure during the Cryo SPD.

This mechanism is predominant in FCC metals with high stacking fault energy materials and HCP materials with a high c/a ratio. This is also predominant in SPD processes where the strain inhomogeneity is more during processing and/or the adiabatic heating causes local softening in the deformation zones.

11.1.2 Grain Refinement by Strain Hardening

The grain refinement and microstructural evolution during SPD is as follows: at low strains, grains are first divided into Cell Blocks (CBs), which contain dislocation cells. Dislocation tangle zones (DTZs) may also develop inside CBs. With increasing strains, CBs may further subdivide into smaller CBs and DTZs may transform into dislocations cells. Sub-grains will develop from both CBs and dislocations cells. The latter become sub-grains when the misorientation across their boundaries are so large that they develop their own unique slip systems. The misorientation across sub-grain boundaries increases with further strain and eventually becomes large enough to transform the sub-grain boundaries into low-angle GBs or high-angle GBs, as shown in Figure 11.2.

The substructure developed during the SPD process resembles the idealised models in the initial stages as discussed in Chapter 1. The area fraction of LAGBs is considerably higher in initial passes, and the tendency of combining two low-angle grain boundaries to form the boundaries of large misorientation becomes greater at higher strains.

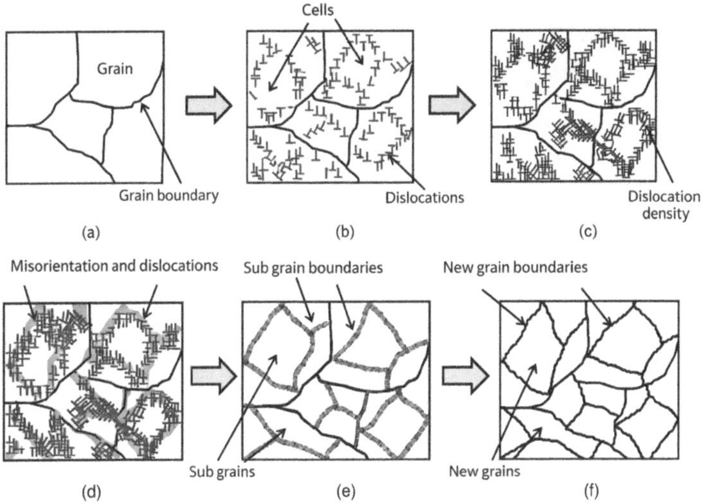

FIGURE 11.2 Schematic representation of grain refinement mechanism: (a) typical microstructure before SPD, (b) initiation of dislocation movements, (c) increased dislocation densities, (d) misorientation at the dislocation cell walls, (e) development of sub-grain boundaries and (f) new grains (with low-angle or high-angle grain boundary).

Also, the merger of two low-angle boundaries, forming a high-angle boundary, always results in a net decrease in the total energy of the interface. The vital aspect of the HAGBs is that they can function as an effective dislocation source, which influences the mechanical properties to a greater extent.

11.1.3 Mechanism of Grain Refinement in Composites

Figure 11.3 shows a schematic illustration of the grain refinement mechanism in composites. As illustrated, SPD techniques, including FSP promote particle fragmentation and dislocation pile-up. The presence of particles hinders dislocation movement, resulting in the accumulation of dislocations in localised regions and expediting the process of dynamic recrystallisation. Furthermore, during the subsequent growth of recrystallised grains, the presence of particles in close proximity to grain boundaries leads to the phenomenon of Zener pinning, which restricts the motion of the grain boundary. The pinning force exerted by these particles is inversely proportional to their average size. The pinning force swiftly counterbalances the curvature and dislocation driving forces, effectively impeding the growth of recrystallised grains, ultimately resulting in a nano-sized microstructure.

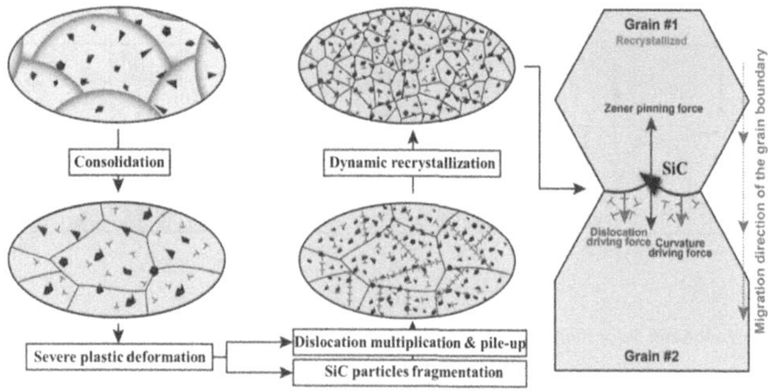

FIGURE 11.3 Schematic illustration of grain refinement mechanism.

11.1.4 Effect of Plastic Deformation on Grain Orientation and Texture

A material undergoing severe plastic deformation (SPD) will develop a preferred orientation or texture, in which certain crystallographic planes tend to orient themselves in a preferred manner with respect to the direction of maximum strain. The tendency for slip planes in single crystal to rotate parallel to the axis of principal strain but in polycrystalline material complex interaction between multiple slip system makes texture more complex. Since the individual grains in polycrystalline material cannot rotate freely, lattice bending and fragmentation will occur. The simplest texture produced by deformation is fibre texture or mechanical fibering, which is brought about by the alignment of inclusions, cavities and second-phase constituents in the main direction of mechanical working. It is similar to the arrangement in naturally fibrous materials. The other texture is crystallographic fibering or crystallographic texture, which is due to crystallographic reorientation of the grains during deformation. Both textures are important factors in producing the mechanical properties of plastically worked materials.

In an ideal deformation texture, a definite crystallographic direction lies parallel to the loading axis, and the texture is symmetrical around the loading axis. BCC metals have fibre texture with <110> direction parallel to loading axis. FCC metals can have double fibre texture with both <111> and <110> directions parallel to the loading axis. The <111> texture is favoured by easy cross slip and predominates in metals such as aluminium. In hcp metals, the basal plane rotates so that the <$10\bar{1}0$> direction coincides with the loading axis. The texture resulting from plastic deformation is strongly dependent on the slip and twinning system available for deformation, but it is not generally affected by processing variables. The most important mechanical variable is the geometry of flow and the amount of strain imposed.

The recrystallisation of cold-worked metal generally produces a preferred orientation which is different from and stronger than deformation texture. This is called annealing texture or recrystallisation texture. An outstanding example is the cubic texture in copper where {110} plane lies parallel to the rolling plane <110> direction parallel to the direction of rolling. Since the existence of a recrystallised texture depends on a preferential orientation of the nuclei of the recrystallised grains, the resulting texture is strongly dependent on the texture produced by the deformation. Another important variable that affects the annealing texture are the composition, the initial grain size and grain orientation of the alloy, and the annealing temperature and time. Generally, the factors that favour the deformation of fine recrystallised grains also favour the formation of an essentially random orientation of recrystallised grains. Moderate cold reduction and low annealing temperatures are beneficial.

The texture due to severe plastic deformation results in crystallographic anisotropy in the polycrystalline materials called crystallographic anisotropy. The strength of single crystal is highly anisotropic. The crystallographic anisotropy due to texture aims to approach the anisotropy of single crystal. The yield strength and, to a lesser extent, tensile strength are the properties most affected. The yield strength in the direction perpendicular to the main (longitudinal) direction of working may be greater or less than the yield strength in the longitudinal direction, depending on the type of preferred orientation existing. This type of anisotropy is most frequently found in non-ferrous metals when they are severely deformed. Crystallographic anisotropy can be eliminated by recrystallisation, although the formation of a recrystallisation texture can cause reappearance of a different type of anisotropy. Crystallographic anisotropy may also result in the elliptical deformation of a tensile specimen.

Texture evolution in ECAP-ed materials studied by one in the author's group on Al 2014 processed in all routes up to five passes in a die with a channel angle of $\Phi = 90°$ and a corner angle of $\psi = 20°$:

Al 2014 alloy ECAP-ed in (i) annealed (ii) solutionised and (iii) solutionised+aged in all routes.

The textures evolution during ECAP of 2014 Al alloy strongly depends on the starting microstructure and processing routes of ECAP. The prominent texture components are $A_E/\bar{A}_E$ and $B_E/\bar{B}_E$ in addition to several weaker components in all four processing routes. The overall texture evolution is stronger in Route A in solutionised condition compared to other conditions and processing routes.

The strong texture evolution in solutionised condition has been attributed to higher strain hardening of the matrix due to a higher amount of solute. In the annealed as well as solutionised plus aged condition, weak texture evolution is observed due to strain scattering during extensive precipitate fragmentation in solutionised+aged condition and due to the dissolution of precipitate in the annealed condition during ECAP. The heterogeneity in texture evolution is less in Route C compared to the other routes. It is attributed to the reversal of shear. In Route A, texture heterogeneity is more than in the other Routes B_C and B_A because of monotonic deformation. In Routes B_C and B_A, the texture evolution in the outer region near the surface of the billet changes the orientation as the number of passes increases, possibly creating a stronger texture evolution different from the centre.

11.2 MECHANICAL PROPERTIES

In the SPD process, the tensile strength of the material increases, and there is a decrease in the ductility in every cycle, and also an abrupt decrease in the elongation. As the specimen is deformed and the strain is induced in the material, the mechanical properties change. Dislocations play an important role in determining the mechanical behaviour of the materials. The plastic deformation in the specimen corresponds to the motion of dislocations and thus, the ability of a material to deform depends on the ability of the dislocations to move. This reveals that restricting the motion of dislocations makes the material stronger and harder because greater strength is required to initiate deformation. All the strengthening mechanisms, such as strain hardening, are based on this principle of hindering the dislocation motion. Increase in dislocation (Figure 11.4) density implies that the dislocations are positioned more closely. On average, strain interactions between dislocations are repulsive in nature, and the net result is that the motion of a dislocation is hindered due to the presence of other dislocations. This hindrance gets increased with the increase in dislocation density, and thus, higher stresses are required to deform the material.

It is observed that the dislocation density gets increased, but the rate of increment gets decreased with the increase in accumulated strain. This is due to two mechanisms: strain hardening and dynamic recovery. At small values of strain, the strain hardening mechanism is dominant and dislocation density increases drastically. With the

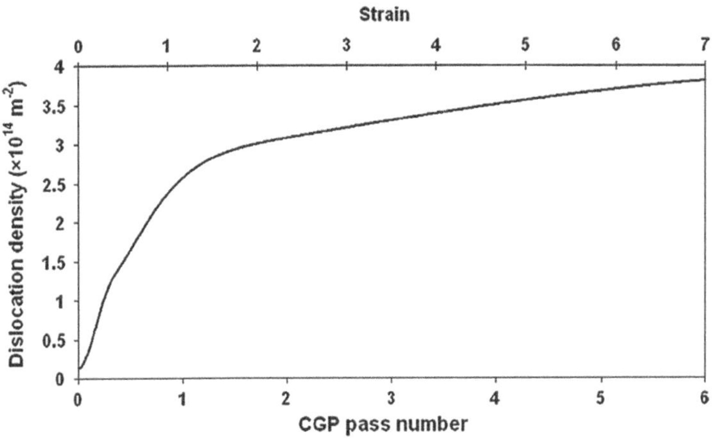

FIGURE 11.4 Evolution of dislocation density during SPD (schematic).

increase in dislocation density, the driving force for dynamic recovery gets increased. The role of dynamic recovery is dislocation annihilation, and so the rate of dislocation density increment gets decreased. The trend is generally observed in the materials processed by all SPD techniques.

11.2.1 Strength and Ductility

The general trend in the variation of strength of the SPD samples is that the strength increases till a certain number of passes (the number is different for different materials) and then decreases for further passes. The increase in strength is quite significant after the first pass. The trend is much similar to dislocation density as shown in Figure 11.4. However, all the studies on SPD show a decrease in ductility, and the decrease is found to be quite drastic after the first pass. Yield strength (YS) and ultimate tensile strength (UTS) increased greatly after the first pass and then decreased marginally in the later passes. The yield ratio (YS/UTS) indicates that strain hardening takes place negligibly in all the SPD samples, while the undeformed specimens exhibit extensive strain hardening behaviour. Ductility (elongation to failure) decreased significantly after the first pass and increased slightly in the later passes. However, the change in ductility in the later passes is observed to be different for different materials. Figure 11.5 schematically shows the tensile properties of the SPD process in general.

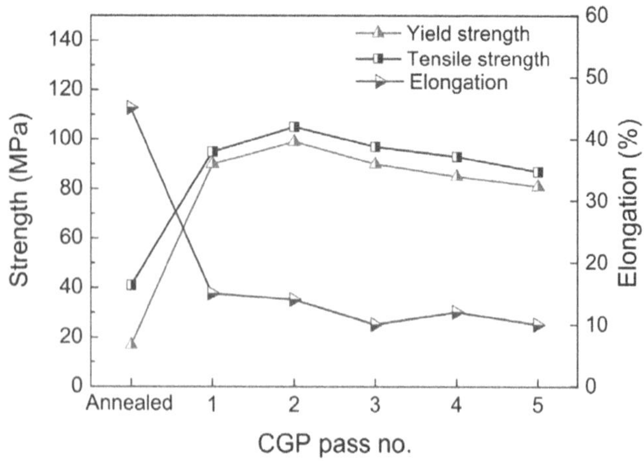

FIGURE 11.5 Tensile properties of SPD with pass number.

The significant increase in the first pass is because of the grain refinement phenomenon that occurs due to severe deformation and also due to the increase in dislocation density, which requires higher applied stress for dislocation motion. Marginal increase is observed in YS and UTS after the second pass, and both of them reached their maximum. This is due to the increased annihilation of dislocations with the increase in accumulated strain and formation of micro-cracks. Elongation decreased significantly after the first pass. The drastic decrease in ductility is because of the decrease in strain hardenability. The drastic decrease in ductility is due to loss of strain hardening ability, which is given by the following equation:

$$H_c = \left(\frac{\sigma_{UTS}}{\sigma_{YS}}\right) - 1$$

where H_C is strain hardening ability, σ_{UTS} is ultimate tensile strength and σ_{YS} is yield strength.

This behaviour of SPD materials displaying negligible strain hardening is explained mechanistically as a balance between the rate of dislocation generation and the spreading rate of trapped lattice dislocations (TLDs) at grain boundaries. One reason for negligible strain hardening is that the mean free dislocation length at the corresponding stress level is comparable to the grain microstructures. Dislocation tangling, causing strain hardening, is unlikely to occur in this grain size condition. Another reported reason is that, due to the relatively fast TLDs spreading kinetics, dynamic recovery takes place during tensile deformation, causing a negligible increase in the lattice dislocation density. However, ductility increases after a certain number of passes which variation depends on the material and its composition. This is because of the increase in strain hardening ability, which is due to the dominance of dynamic recovery of dislocations at higher strains.

In the case of precipitation hardenable alloys, the increase in strength after the first pass is much higher than that of single-phase or non-precipitation hardenable alloys. This is due to the following reason: precipitation hardenable alloys lead to the formation of fine dispersoids. With the increase in strain, the average misorientation of formed cells increases more rapidly than that of single-phase alloy and hence, the rate of strength improvement increase in the case of precipitation hardenable alloys, and a high strength can be obtained within a fewer number of passes. However, in precipitation hardenable alloys after few passes, a drop in strength is observed and the reasons given for this phenomenon are: coarsening of precipitates, flow-softening and micro-cracking. However, the ductility of precipitation hardenable alloy is higher than that of pure metal or non-precipitation hardenable alloy. Higher ductility in the case of precipitation hardenable alloy is because of the presence of precipitates.

11.2.2 Hardness Behaviour and Homogeneity

Hardness and strength are related properties, but they are not exactly the same. A similar trend to that of strength variation is expected for hardness, but in most cases, it is not similar. Hardness increased significantly after the first pass and decreased marginally in the later passes. Improvement in hardness is due to the rise in dislocation density and work hardening of the material. The decrease in hardness is due to coarser grain size and the near dislocation-free nature of the grains that occur at larger strains. Another reason for the decrease in hardness is dynamic recovery which is activated at higher strains like strength and dislocation density. Generally, strength-hardness behaviour of SPD materials does not obey the empirical relationship between hardness and strength derived for coarse-grained materials. This phenomenon is explained by the fact that hardness is a surface phenomenon and is defined as resistance of the material to indentation. On the other hand, tensile strength represents the inherent strength of the material and is defined as resistance to deformation. Another investigation has also indicated the same phenomenon that severe plastically deformed materials do not obey this relationship. This is because YS corresponds to the onset of plastic deformation around mean strain less than 1%, but hardness corresponds to a mean strain of about 10% due to the differences in the work hardening abilities. However, there can be exceptions for some materials and processes, where this strength-hardness relationship is obeyed.

Hardness behaviour and homogeneity of SPD materials are analysed by measuring the hardness values along the transverse cross-section of the samples. Homogeneity is entirely dependent on the SPD process and surface friction conditions. Hardness behaviour and its homogeneity both in centre and surface increased with the increased number of passes. This is due to the lesser effect of strain magnitude on microstructural changes during the higher number of passes. In order to investigate the uniformity of hardness distributions, the inhomogeneity factor (I.F.) is used. This is calculated by using the following formula:

$$\text{I.F} = \frac{\sqrt{\sum_{i=1}^{n} (h_i - H)^2 / (n-1)}}{H} * 100$$

where n = number of hardness measurements, h_i = hardness value of i-th measurement and H = mean hardness value. The lesser the I.F. value, the higher the homogeneity. Normally, the I.F. value will be less in the pre-deformed condition, which increased significantly after the first pass, and then decreased gradually for further passes. Another observation is that, for every pass, the I.F. value of the surface is lower than that of the inner, which means the inner section is more homogeneous compared to the surface. This can be related to the pure shear deformation that occurs in the inner section and the existence of an additional one in the surface due to the friction effect, i.e., the interface of surface and the die.

11.2.3 Creep Properties

Deformation of materials over time when subjected to constant stress and temperature is called as creep. The creep deformation is due to various creep mechanisms. The applied stress and temperature significantly influence the dominant creep mechanism and the role of grain boundaries in the deformation process. The mechanisms include (i) Climb-assisted glide, where dislocations move around obstacles, (ii) Thermally activated glide which involves cross slip, (iii) Coble creep, where the grains shift perpendicular to the stress, (iv) Diffusion creep where vacancies diffuse through the crystal lattice of crystalline solids and (v) Dislocation creep, normally occurs at low temperatures, where shear stress drives diffusion into the dislocation core. Thus, the creep properties (resistance) depend on external parameters such as applied stress and temperature and properties of the materials, especially the arrangement of grains.

In case of SPD materials, in general, creep occurs by the movement of dislocations and grain boundary sliding. The creep resistance increases considerably after one pass; however, successive SPD pressings lead to a noticeable decrease in the creep properties. Increase in the number of passes results in more fine grains and due to increase in grain boundary area, grain boundary sliding dominates the dislocation process. Besides, the angle of grain boundaries has a significant impact on creep behaviour. SPD materials containing high-angle grain boundaries generally exhibit greater creep resistance compared to those with low-angle grain boundaries since it is more difficult for dislocations to move across the boundary, thus hindering creep deformation. However, when grain boundary sliding is dominating the creep mechanism, creep of SPD materials is accelerated if it has high volume fraction of high-angle grain boundaries. The role of grain boundary sliding in creep considerably differs in both SPD and coarse-grained materials.

Besides, SPD samples after creep exposure have mesoscopic shear bands exceeding considerably the average grain size. The formation of these bands could be influenced by inhomogeneity of microstructure created during SPD. The cavities and micro-cracks can nucleate and propagate along the mesoscopic shear bands. The evolution of creep damage and fracture of ultra-fine-grained specimens could be influenced by synergetic effects of grain boundary sliding, formation of shear bands which were oriented along the direction of the last SPD pass and cavitation. Achieving homogeneous microstructure is most important during SPD to have better creep resistance. The creep behaviour also depends on texture of the material processed by SPD.

Diffusion creep, which is dominant at high temperatures, is not important in SPD materials because the ultra-fine grains are unstable at these temperatures and the grains grow sufficiently to preclude any significant contribution from Coble diffusion creep. The potential for using SPD materials for high temperature creep is due to the fact that the ultra-fine grains are reasonably stable at elevated temperatures. The SPD process as well as the process variables should be selected carefully so that the ultra-fine-grained materials can withstand the dominant creep mechanism at the service temperature and stress.

11.2.4 Fatigue Properties

Fatigue behaviour is based on metal/alloy characteristics, process parameters of SPD, operation environment and the deformation behaviour during the fatigue loading. In general, the fatigue limit increases significantly as the grain size decreases, and the ratio between fatigue limit and UTS also increases. The higher density of grain boundaries due to SPD are effective barriers to dislocation movement during cyclic loading. Dislocations must navigate these boundaries to propagate deformation, with each boundary acting as a pinning point that impedes their movement. This increased barrier to dislocation movement effectively delays the onset of crack initiation due to slip plane extrusion and intrusion mechanisms. Fatigue is not only associated with the formation of specific intergranular dislocation structure but likely to be connected with grain boundaries and its characteristics. Fatigue behaviour differs in load controlled cyclic (high-cycle fatigue) and constant strain amplitude (low-cycle fatigue) conditions. The fatigue behaviour also depends on the material parameters. HCP metals such as CP-Ti after SPD are stable after fatigue as against FCC metals such as Cu and Al, which showed significant grain growth. Even under the strain amplitude mode, ultra-fine-grained HCP metals resist from softening even at fairly large plastic strain amplitudes. Softening is generally observed in FCC metals. The slip occurs only in the prismatic plane in the case of HCP and hence the substructure developed during SPD cannot be relaxed by cross slip. This may be the reason for resisting softening in HCP. Whereas cyclic softening in FCC metals is due to grain growth in the localised shear bands. Alloying HCP metals such as Mg can increase the fatigue performance in aluminium alloys by restricting the dislocation mobility process with solute atoms.

Grain boundary precipitation phases and the width of the precipitate-free zone (PFZ) play a critical role in influencing fatigue properties. Also, the presence of more high-angle grain boundaries favours micro-crack nucleation during fatigue loading. Fragmentation of the second phase precipitates due to heavy straining improves fatigue life by avoiding the initiation of micro-cracks. For example, the presence of Mg_2Si, in the aluminium matrix, leads to a significant improvement in the low-cycle fatigue life of Al 6063 due to fine and fragmented particles after processing with SPD. SPD technique and the process parameters should take care of these aspects. For example, SPD of Al 6063 in 90° ECAP die shows better fatigue resistance compared to samples processed in 120° die. This is due to higher degree of grain refinement in 90° die. Besides homogeneity in grain size also affects the fatigue properties. The fatigue strength is an intricate balance between grain refinement and strain distribution. The presence of compressive residual stresses and the associated grain refinement significantly enhance the fatigue resistance of the SPD materials.

It is a known fact that SPD increases the plastic strain energy in the materials. Accumulation of plastic strain energy is critical in causing damage and fracturing materials when it reaches a critical threshold and is essential for evaluating the fatigue life of metals. Critical value is typically determined through uniaxial tension tests, although the plastic strain energy densities from uniaxial tension and cyclic loading differ. A higher number of passes, i.e., more strain induced, causes more plastic strain energy and is likely to reduce the fatigue life in SPD materials. The major factor that influences fatigue life is the surface finish. The SPD processes should aim to have a fine surface finish when subject to fatigue loading. The other factor is the notch sensitivity. Generally, notch sensitivity is higher for ultra-fine-grained material as compared to coarse-grained counterpart. The above should be considered while processing the material in SPD for further applications involving fatigue loading.

11.3 FABRICATION PROPERTIES

11.3.1 Strain Rate Sensitivity and Superplasticity

In the SPD materials, the strength increases, and the ductility of the sheets decreases with the increase in strain rate. The value of strain rate sensitivity (SRS) increases with the increase in strain or after a number of passes during SPD process. This is due to the reduction of grain size with the increase in the number of passes. Another explanation for the enhancement of SRS is the mechanism linked with the grain boundary, such as sliding or shear of grain boundaries, which is accompanied with atomic diffusion and relaxation. The rate of increment in SRS decreased with the increase in strain. This is because, with the increase in strain, annihilation of dislocations occurs due to dynamic recovery. This phenomenon limits the minimum grain size that can be achieved by severe straining. Thus, the value of SRS will not be greater than a specific value. After a few passes, a drop in SRS is observed. This is due to the formation of micro-cracks. SPD increases the SRS and decreases the strain hardening ability of materials. The increase in SRS and fine grains will help to increase the superplasticity of the materials.

11.3.2 Deep Drawability and Workability

Deep drawability is used to define its ability to attain a large amount of plastic flow in the plane of the sheet metal and provide adequate resistance to through-thickness flow. This property is called the normal anisotropy (r_m). It is calculated in terms of value using the relationship, $r_m=(r_0+2r_{45}+r_{90})/4$, where the subscripts denote angles between the longitudinal axes of the tensile specimens and the rolling direction (RD) of the sheet. Each r value is determined by the strain ratio (width strain/thickness strain) using a simple tensile test. The deep drawing sheet must possess high normal anisotropy (r_m value) and low Δr $((r_0+r_{90}-r_{45})/2)$.

The SPD processes, especially CGP and ARB tend to develop texture and thereby create anisotropy in the sheets, by inducing a shear texture favouring higher r values in sheets. In general, r value gets increased especially along 45^0 to the rolling direction or loading direction (RD). The maximum value of r is obtained along the 45° to the RD for the homogenous materials. A decrease in *r* value is observed for the materials along the 90° to the RD. Higher r_m value and low Δr value increase the draw ratio and deep drawability.

In case of rods and other shapes, workability is a measure of its capability to form required shape by forming. Workability can be defined as limiting plastic deformation beyond which fracture occurs in the workpiece. Workability is a function of the material as well as a function of the process parameters. However, the degree of shape change is limited by the capacity of the workpiece to deform plastically. Workability determines the suitability of materials produced by SPD to industrial processing. For determining workability, upsetting tests were carried out on specially designed hexagonal collar specimens machined from specimens processed by SPD. The collar specimens were developed to create a mode of deformation dominated by the circumferential tensile stress on the periphery of the collar. Illustration of the collar specimen is shown in Figure 11.6 where hexagonal collar (d_0), height of the sample (h), height of collar (z) and diameter of the sample (d) and preferred dimensions of the specimens used are of different heights to diameter ratios of 0.75, 1.0 and 1.5.

The specimens are upset between well lubricated flat platens and carried out in stages until visible sign of cracking at the outer-barreled surface is observed. The fracture strain is calculated from the final dimensions of the collar. The axial strain is calculated from the initial and final height of the collar. The hoop strain is calculated from the average diameter of three different corners of the hexagonal collar

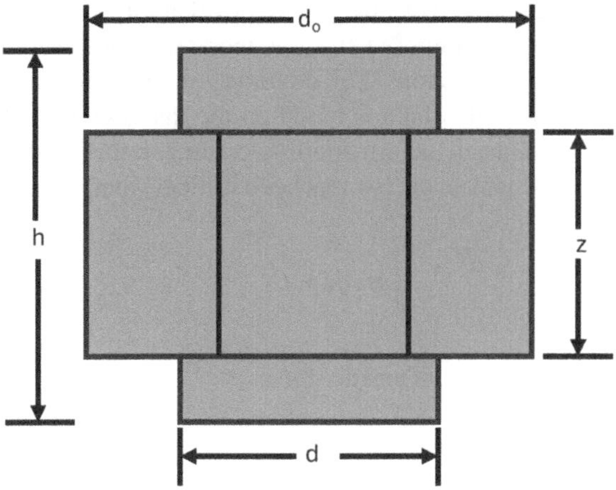

FIGURE 11.6 Geometry of collar specimens used for workability deter mination.

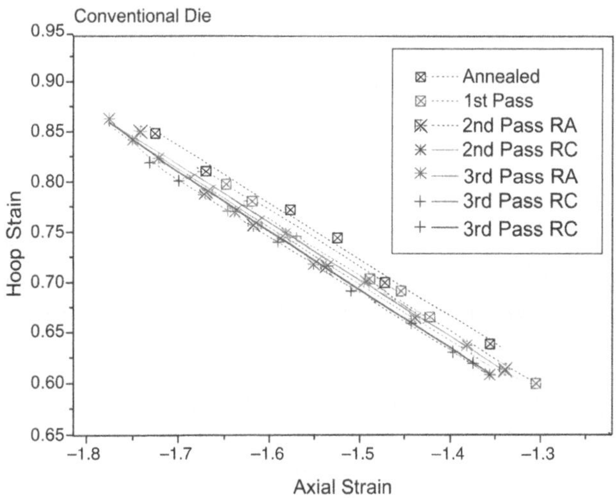

FIGURE 11.7 Workability limits diagram for ECAP-processed specimens.

initially and after fracture. Workability limit diagrams are plotted between axial strain and hoop strain as shown in Figure 11.7, plotted for ECAP-ed CP-Al carried out in the author's laboratory.

For the first pass ECAP, workability slight reduced as compared with the base material. The results also indicate that Route C has better workability than Route A. In general, workability decreases as the number of passes or imposed strain is increased. However, the workability limits are very close. The other way to evaluate workability is the Cockcroft model, which is based on the macroscopic role of tensile stresses and plastic deformation. According to this model, fracture occurs when the tensile energy reaches a critical value.

$$\int_{0}^{\bar{\varepsilon}_f} \sigma_\theta \, d\bar{\varepsilon} = C_1$$

where σ_0 is the maximum tensile stress

ε - Effective strain
ε_f - Effective fracture strain
C_1 - Cockcroft and Latham constant.

The Cockcroft model also indicate same as that of the workability limit diagrams.

As a conclusion, it can be stated that though SPD decreases the workability, the material can be worked if the material shows sufficient ductility even after SPD. Shaping components using SPD material is feasible.

11.3.3 Machinability

Machinability refers to the ease or difficulty with which a material can be cut and shaped during a machining process, influencing factors like tool life, surface finish, and required power. Apart from the machining parameters such as geometry of the tool, tool material, cutting speed, feed rate, and coolant, material property such as hardness, strength, ductility, and microstructure of the workpiece material greatly affect machinability. In general, high hardness and strength lead to poorer machinability. SPD processes mainly aim to increase strength and hardness by refining the microstructure. SPD materials have an effect on machining to adopt new conditions and strategies. Manufacturers are facing the advantages and disadvantages of SPD materials. The machinability of materials significantly changes when the material undergoes severe plastic deformation. The influence of machining parameters, such as cutting speed, feed and depth of cut on the cutting forces needs to be investigated due to change in materials properties due to SPD.

The machinability of Grade 2 Titanium alloy decreases when processed by ECAP. The main cutting force required for turning these specimens reached the same level of Ti-6Al-4V alloy, and is much larger than that required for the coarse-grained Grade 2 Titanium. Attrition is found to be the main wear mechanism acting on the Grade 2 Titanium, regardless of its condition, since a strong adhesion of workpiece material followed by tool particles pull out is observed. However, in the case of CP-Al, successive ECAP processed specimen, better surface finish is observed due to the hardness enhancement. The machinability is enhanced in aluminium in contrast to titanium. In general, the SPD plays a complex role in machinability and is more influenced by the machinability of the base material rather than the SPD.

11.3.4 Weldability

In fusion welding, the parent material melts and hence the properties achieved by SPD can no longer exist. In brazing and soldering, the parent materials experience high heat, though it does not melt. The retention of the properties achieved by SPD depends on the thermal stability of the materials. In case of brazing or soldering, the properties of the components depend on the brazing/soldering rods and so, the base material property is not taken into account. The weldability of SPD materials deals mainly with solid state welding processes such as spot welding, friction welding etc. The increase in electrical resistivity due to SPD improves weldability by increasing heat input and so lower welding parameters are sufficient for spot welding. The decrease in thermal conductivity is useful in friction welding.

It is well known that friction welding results in achieving the fine grains in the welded zone, and this change provides better strength to welding interface and most often the failure occurs in the base material since the thermomechanical affected zone and the heat affected zone comparatively have better strength due to grain refinement and dislocation density offered by the friction welding process. Since the SPD process increases the strength of the base material, the strength of the friction-welded joints is better than the strength obtained by friction welding of their coarse-grained counterpart.

Friction welding studies on SPD materials were conducted by one of the authors' research group. ECAP is selected for fabricating SPD rods of structural steel (IS 2062) for friction welding. The ECAP is carried out in room temperature (RT-ECAP) and at 550°C (H-ECAP) adiabatically. The joining of ECAP-ed samples is approached in two different ways and means: (i) Friction welding of ECAP-ed samples and (ii) ECAP of Friction-welded samples.

11.3.4.1 Friction Welding of ECAP-ed Samples

The SPD samples done by RT-ECAP and H-ECAP show improved tensile strength as compared to normal structural steel friction-welded in the same condition. The improvement is about 50% in RT-ECAP and the H-ECAP-ed samples while the friction-welded sample strength has improved by 70%. The stress-strain diagram for all the conditions is shown in Figure 11.8. The interesting fact to note is that even after friction welding the structural steel retains almost its strength achieved due to ECAP.

The microhardness survey (Figure 11.9) shows the hardness distribution of the welded samples. In the case of FW in the non-ECAP-ed

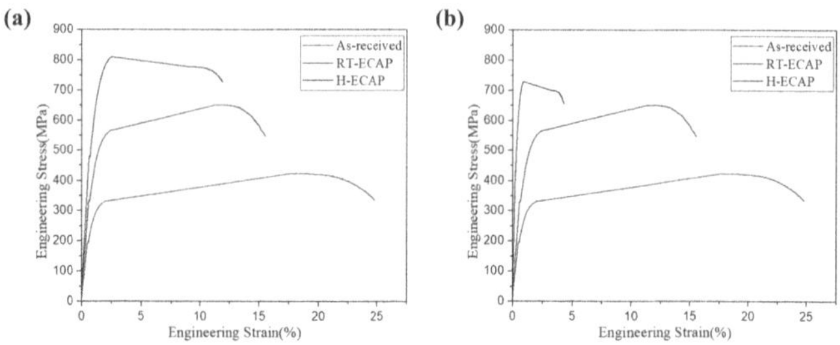

FIGURE 11.8 Stress-strain curves of (a) ECAP-processed and (b) friction-welded samples.

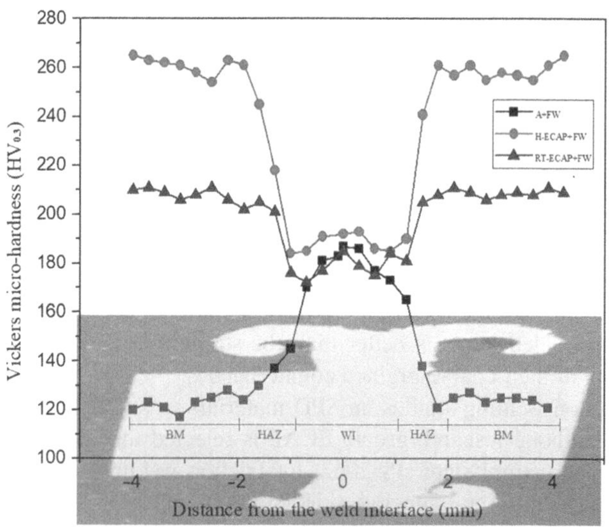

FIGURE 11.9 Vickers's microhardness profile of the welding joints.

condition, the HAZ and WI show higher hardness due to grain refinement achieved by FW. However, in the ECAP-ed conditions, the HAZ and WI show lower hardness than the base material. It implies that the effective grain refinement occurs due to ECAP rather than due to FW. However, the friction-welded ECAP samples show better hardness than the friction-welded non-ECAP-ed samples. The hardness distribution and the tensile strength go in line. The fracture occurs in the base material for friction-welded non-ECAP-ed samples, as opposed to in the HAZ zone in the case of friction-welded ECAP-ed samples.

11.3.4.2 ECAP of Friction Welded Samples

In this section effect of ECAP on friction-welded structural steel IS2062 is discussed. After friction welding, the samples are ECAP-ed at ambient temperature for two passes. The strength has been improved up to 70% by the ECAP process carried out at ambient temperature for the friction-welded structural steel. The microhardness survey shows that the microhardness is almost same in all the regions of the sample due to uniform refinement in all zones of the welded joint.

Figure 11.10a shows the stress-strain curve of as-friction-welded (AW), annealed after friction welding (AA), ECAP first and second pass. It is well understood that after ECAP, friction-welded samples show higher strength than AW samples, mainly due to grain refinement. The interesting part is the hardness in the various zones of the samples (Figure 11.10b). Friction-welded structural steels (AW) show fine grains in HAZ and WI as compared to the base metal, which is expected during the friction welding, which is reflected in the hardness curve across the welded zone. The hardness is higher in HAZ and WI zones. The stress relief annealed friction-welded (AA) samples show almost a similar trend, but the hardness is lower as compared to

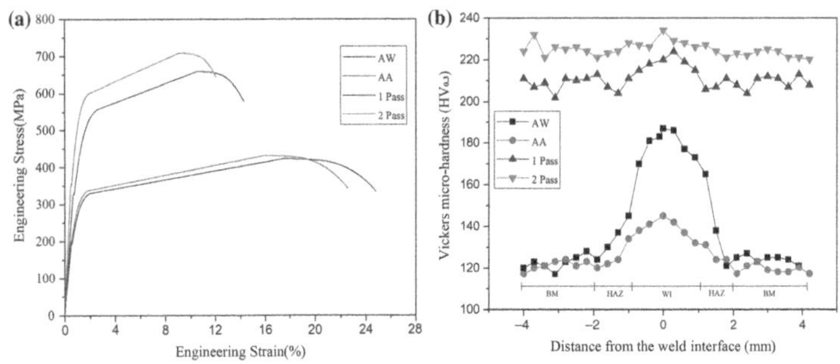

FIGURE 11.10 (a) Engineering stress-strain curve, (b) Vickers microhardness of the FW specimen before and after ECAP pass.

AW samples due to grain growth in the HAZ and WI zones, and the hardness of the base material remains the same. The hardness gradient across the length is reduced. The samples after ECAP show higher hardness than the AW and AA samples in all the regions. It is interesting to note that the hardness across the length is uniform, and the hardness gradient is one. It indicates that ECAP refines the grains in all the regions uniformly, irrespective of initial grain size. This phenomenon is more useful in the design of the welded structures.

11.4 SURFACE PROPERTIES

11.4.1 Wear Resistance

SPD leads improvement not only in the mechanical properties but also in wear resistance, especially in dry abrasion conditions. The wear resistance improved continuously with the increase in the number of passes. Also, the effect of initial passes on improving the wear resistance is more significant compared with the effect of later passes. Friction is an important property that affects the performance and wear resistance in structural members that are sliding against each other. It is reported that ultra-fine-grained low carbon steel obtained by SPD showed low coefficient of friction as compared to the coarse one. Similarly, ultra-fine-grained titanium processed by ECAP (grain size of 0.3 microns) showed low friction coefficient than coarse-grained material at all temperatures (room temperature to 800°C) and at all loads (480–2400 N). The friction coefficient of the material processed by the two different routes (Route B_C and C) is almost equal. The friction coefficient values of coarse-grained titanium increased with increasing temperature while for the fine grained it decreased. SPD of pure Mg. AZ91 alloys processed by ECAP exhibit lower wear rates compared to their untreated counterparts, showing a similar trend. Wear tests have demonstrated that the wear rates can decrease significantly due to the combined effects of grain refinement and improved mechanical properties. The degree of wear resistance enhancement through ECAP also depends on processing parameters such as the number of passes and the specific conditions (temperature, die design).

Wear studies on ECAP-ed pure Mg, Mg-Li alloys and in Al 2014 were conducted by one of the author's research group. The results are briefly discussed. The wear behaviour is studied in terms of wear rate and coefficient of friction (COF) and possible wear mechanisms are analysed. In general, wear resistance is greatly improved after ECAP and the wear rate and COF decreased. In the case of pure magnesium,

Mg-4Li and Mg-9Li alloys, the samples processed in the Route B_C show better grain refinement, superior mechanical and wear resistance as compared to the other processing routes, whereas in Mg-14Li, the sample processed in Route B_A shows better grain refinement, superior mechanical and wear resistance as compared to the other processing routes. Pure Mg and Mg-Li have a crystal structure of HCP (alpha phase), Mg-9 Li has a crystal structure of HCP (alpha) and BCC (beta), whereas Mg-14 Li has a crystal structure of BCC (beta phase).

In case of pure Mg, the average COF decreased after four passes of ECAP in all processing routes. The wear rate decreased by 26% and 37–50% after one pass and four passes of ECAP, respectively. The sample processed in the Route B_C after four passes of ECAP displayed the minimum wear rate as compared to other processing routes. The wear resistance increases with the increasing number of passes. The sample processed in four passes of Route B_C showed the lowest average COF and wear rate owing to its fine grain and highest hardness. Several investigations have classified the different wear mechanisms into two wear regimes, namely, mild and severe, depending on the applied load, sliding velocity and sliding distance. The mild wear regime is further categorised into two sub-regimes namely oxidation wear and a delamination wear. In the present work, the combination of applied load (5 N) and sliding velocity (1 m/s) falls within the mild wear regime where the oxidation wear and delamination wear mechanisms would take place predominantly. The un-ECAP-ed Mg and ECAP-ed Mg in all passes exhibit significant delamination and distinctive adhesive wear scars oriented parallel to the sliding direction. Oxidation wear was confirmed by the prominent oxide peak in the EDS analysis. However, stacks of delamination formed along the sliding direction, with the wear band width being wider in the annealed samples compared to the ECAP-ed. In the ECAP-ed samples, similar wear mechanisms were noted, but both the wear band width and delamination wear were reduced with increasing ECAP passes, especially after four passes. This improvement is linked to increased hardness and enhanced wear resistance. The choice of ECAP processing route also affects wear properties due to differing shear characteristics. Samples processed via Route B_C showed more uniform and equiaxed grains, leading to lower COF and wear rate. These samples exhibited narrow wear scars and less delamination compared to those from other routes.

In the case of ECAP of pure Mg, the wear rate is directly proportional to the hardness obeying Archard's law indicating that the mechanism of wear is unaltered even when the material is subjected to SPD. It implies that under the conditions of wear testing selected in this work, the wear resistance only depends upon grain refinement and the mechanical properties achieved. However, the wear mechanism may change if the wear conditions differ since wear entirely depends on the testing parameters and conditions. The wear behaviour of Mg-4Li alloy, Mg-9Li alloy and Mg-14Li alloy under the conditions selected in this study shows the same trend. Among all alloys, Mg-9Li showed a minimum wear rate and COF due to more refined microstructure and superior mechanical properties.

Similarly, ECAP of AA 2014 in different routes shows that the samples prepared through Route B_C show less volume loss compared to those of Route A samples since the sample processed in Route B_C yields higher hardness than Route A due to cross-hardening. The volume loss and the coefficient of friction of the ECAP samples are found to be gradually reducing with the increase in the number of passes in both routes. In this study, the load selected are higher than those selected for Mg and its alloys. At 10N and 30N, the unprocessed samples show plastic deformation and cracks. The cracks propagation gets arrested after ECAP, whereas adhesive wear particles are observed. Transfer of "Fe" from the counter surface and formation of oxide is observed in AA2014.

11.4.2 Corrosion and Stress Corrosion Cracking

The general corrosion behaviour after SPD is complicated, and it depends on the individual material. ECAP-ed Al-23% Si shows decreased corrosion potential after four passes. Up to four passes, corrosion and pitting susceptibility increased due to the presence of biggest voids resulting from the breakage of the brittle large primary silicon crystals during ECAP. After four passes the voids evanesce and results in the homogenous ultra-fine-grained structure and resulted in improved corrosion resistance. Corrosion resistance of ECAP-ed pure Al in 3.5% NaCl increased as compared to its coarse-grained counterpart and the corrosion resistance increases as the number of passes increases. Apart from grain refinement, the fraction of high-angle grain boundaries increased substantially as the number of passes increases. Higher fraction of high angled and grain boundaries in the microstructure accelerates the passivation process and reducing the intensity of galvanic couple between grain interior and grain boundary, which leads to reduced corrosion rates in ECAP-ed Al. Besides grain size, precipitate size also plays a critical role in corrosion of SPD materials. In SPD samples of anodised Al-Cu alloy, small pits are formed due to fragmentation of precipitates such as Al-Cu-Si-Fe-Mn, Al-Cu-Mg,

Al-Cu-Mg and Al-Cu-Si-Fe-Mn. The pitting corrosion occurs in these precipitates sites and decrease in size of these precipitates improves the corrosion resistance.

Apart from aluminium and its alloys, effect of SPD on alloys like magnesium, copper, steel, etc are also been reported. ECAP of magnesium decreased corrosion resistance and decreased with increase in number of passes. The lower corrosion resistance of ECAP-ed pure Mg is closely related to the strain-induced crystalline defects, including energetic grain boundaries (subgrain boundaries) and dislocations though ECAP-ed Mg forms faster passive film.Fe-20% chromium steel with extremely low carbon and nitrogen shows reduced pitting potential after ECAP. It can be concluded that nobler potentials of ultra-fine grains (UFG) states are realised not only by grain size reduction but also defective deformation-induced UFG.

In case of copper, the corrosion resistance decreases in UFG copper in comparison with the relatively coarse-grained sample. However, the surface of the UFG Copper is rather smooth on a macroscopic scale, and dissolution occurs homogeneously with visible etching. In contrast, the deep and long corrosion grooves are clear in coarse-grained copper. Hence, the corrosion behaviour of UFG copper compared with its coarse-grained counterpart is characterised by two factors: (i) approximately 10% higher average dissolution rate under the steady-state conditions and (ii) a much lower localisation of corrosion. While the first factor is disappointing for engineering applications, the second can be considered as a certain advantage of SPD copper materials. A similar trend is observed in the Ni system as well. However, in case of titanium, ECAP-ed in B_C Route showed better corrosion resistance than coarse-grained Ti (grain size 7 microns) in HCl and H_2SO_4 of concentration from 1 to 5 M. The corrosion in Ti is mainly due to the segregation of impurities (Al and Fe) at the grain boundaries. The concentration of segregation at grain boundaries is less in ECAP-ed Ti due to more area of grain boundaries as compared to coarse-grained titanium. In the same way, ECAP-ed duplex stainless steel shows better corrosion resistance than coarse-grained.

The effect of corrosion on ECAP was studied by one of the authors' groups on different grades of aluminium alloys which are summarised below:

The effect of SPD on Al 2014 processed through different routes of ECAP in 0.35% NaCl is investigated using a potentiodynamic polarisation test, and it was compared with base Al 2014. The corrosion rate decreases initially (after one pass), but in subsequent passes, the corrosion rate increases linearly in all samples processed in all routes except

Route B_A. The corrosion rate processed in Route B_A is almost half of the base material. An important factor that can affect the pitting corrosion behaviour of aluminium copper alloy is the presence of $CuAl_2$ (θ) precipitates. Since the precipitates are bigger in case of base, there is a formation of a large cathode and a small anode mechanism, which increases the susceptibility to corrosion. Due to the reduction in size of precipitates after ECAP cathodic sites become smaller hence corrosion rates get decreased after one pass of ECAP by the small cathode and large anode mechanism. Particularly in Route B_A after third pass the precipitate size is reduced to approximately 40±5 nm and hence the corrosion rate is decreased. The other factor is the grain size. As the grain size decreases, although the formation of passive film is faster in NaCl medium, the Cl^- ions break the passive film, and the corrosion becomes severe as the number of grain boundaries increases. However, in Route B_A though the grain size is finer, it shows more resistance to corrosion as compared to the base material. In a similar manner, ECAP-ed Al-5083 alloys are subjected to corrosion in 3.5% NaCl solution. The sample processed through Route C fourth pass showed better corrosion resistance than Route A fourth pass. The poor corrosion resistance of Route A may be attributed to the fine grains as compared to Route C and may be due to the appearance of new Al–Mg precipitate in Route A. It implies that, the orientation of grains and texture also plays a role in corrosion resistance, which needs further investigation.

Stress corrosion cracking (SCC) depends on various factors such as residual stresses, chemical composition, microstructure and deformation mode. Improving the SCC susceptibility through the utilisation of SPD techniques has rarely been explored. The SCC behaviour of welded joints of AA2195 studied after using laser peening obtained a better resistance to local corrosion compared to samples without peening. Welded joints of AA7075 have also been tested for SCC susceptibility after applying superficial SPD by laser shock peening (LSP) and report decrement of stress corrosion sensitive coefficient compared to samples without LSP. All the peening techniques are surface SPD processes.

To date, there are limited reports of the study of the SCC mechanism when bulk SPD processes are applied. SCC occurs preferentially through the grain boundaries. In general, some degree of convergence of cracks is seen in the SPD-processed materials. SPD materials show high pitting due to the corrosive process. These pits may serve as initiators of cracking. The samples after four passes show lesser cracks as compared to first pass and undeformed sheets when the applied stress is equal to YS of the material. However, the change in the number of cracks between the passes is near the range of the standard deviation estimated during the quantification with slight tendency towards decreasing the number of cracks. As a known fact, the mixture of coarse and fine grains by a bimodal distribution is not preferred for SCC or in other words, the SCC will be higher if I.F. is high. It is a known fact in most of the SPD processes that, as the number of passes increases, the homogeneity is improved. The low cracking on the sample with a higher number of passes is related to the particular widened unimodal grain distribution and the random misorientation distribution obtained from the last deformation pass. Therefore, the particular unimodal and widened distribution might be responsible for the lower electrochemical heterogeneity. Additionally, the random crystallographic orientation could have avoided the presence of significant potential gradients in the sample. This phenomenon could generate a drop in the SCC susceptibility. The other factor that induces the SCC is hydrogen-induced cracking and is due to the weakening of grain boundaries through the absorbance of atomic hydrogen. The amount of hydrogen sites controls the hydrogen mechanism. As the number of passes increases, different structural defects like dislocations, vacancies, as well as local stress concentrators can act as hydrogen traps as against the grain boundaries. This phenomenon decreases the concentration of hydrogen in the grain boundaries. It can be concluded that the SCC susceptibility slightly decreases with the number of passes.

SCC behaviour of Al-7075 before and after ECAP was studied by one in the authors' group using slow strain rate testing (SSRT) method in 3.5% NaCl solution at a strain rate of 10^{-4}/s. The SCC susceptibility of the alloy was evaluated by parameters such as elongation, reduction in area and UTS. The load displacement curves are shown in Figure 11.11. UTS and YS of ECAP-ed sample after SSRT were higher than the UTS and YS of the base material. This gives positive hope.

11.5 OTHER PROPERTIES

11.5.1 Bake Hardenability

Bake hardening is a special property of low carbon steels. The bake hardenability is greater for ultra-fine grain specimens when compared to their coarse-grain counterparts. In case of coarse-grained low carbon

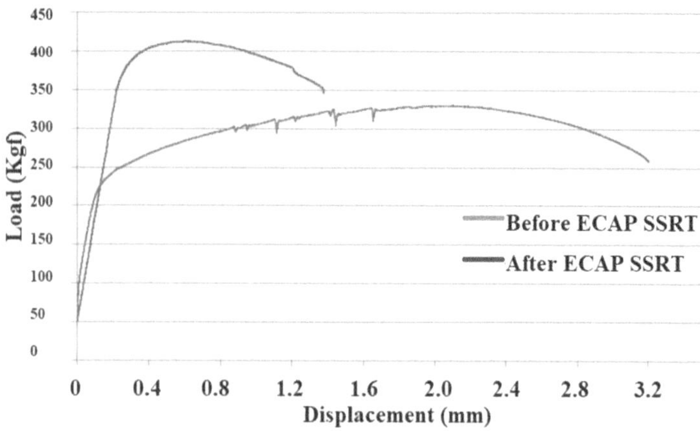

FIGURE 11.11 SSRT load displacement curves of Al-7075 alloy before and after ECAP.

steel an increase in bake hardening of 108% is noticed in contrast to 170% in ultra-fine-grained low carbon steels achieved by SPD.

The UFG materials have greater bake hardenability due to four mechanisms:

i. strengthening due to the Hall–Petch equation, because of both grain refinement to ultra-fine grain size and the increase in the volume fraction of grain boundaries,
ii. higher dislocation density and higher work hardening compared to coarse-grain structure,
iii. the enhanced heterogeneous nucleation sites for precipitates during bake hardening; this is because the applied SPD provides higher density of dislocations acting as preferential nucleation sites for precipitation, and
iv. the increase in the solid solubility of elements due to SPD.

11.5.2 Electrical Resistivity

SPD has a significant influence on electrical resistivity. The electrical resistivity increases as the number of passes increases. After one pass of SPD, the electrical resistivity increases rapidly. This trend can be interpreted by considering the effects of grain boundaries and dislocations on electron scattering. Because of inducing severe deformation to the specimens, the dislocation density increases, and the coarse-grained microstructure becomes refined to a nanostructure. Refining the microstructure leads to an increase in the total surface area of all grains. Electrical resistivity is directly proportional to the grain surface and the number of dislocations, and hence it increases. The rate of increase in electrical resistivity increment decreases after one pass. This is attributed to the effect of dynamic recovery on dislocation annihilation and the decrease in grain refinement rate at large strains. Interestingly, there is a considerable amount of increase in electrical resistivity after the final pass. This could be due to the following reason: deformation through the final pass must have generated some micro-cracks on the sheet surface and the air gaps between the cracks can act as obstacles for the transport of electrons leading to the increase in electrical resistivity.

According to the fast developments in the electronic industries, there is a great demand for high-strength and high-conductivity materials. The alloyed copper is the most widely used material in this respect. The electrical conductivity of these materials is inherently lower than that of their unalloyed counterpart. To resolve this problem, UFG or nanostructured commercial pure copper is used. This goal can be obtained by SPD processes. The conductivity of samples decreased with increasing the number of SPD passes and the strength increases. It is noted that the electrical conductivity of about 81% of IACS with UTS of about 300 MPa can be achieved by SPD, which is much more than copper alloys. Consequently, the electrical conductivity of the pure copper is little influenced by the formation of the UFG microstructures through SPD. Pure copper can be strengthened by SPD with a little reduction in electrical conductivity. Reduction in elongation limits some electrical applications of SPD pure copper but copper with high strength and high conductivity, regardless of elongation, can be used in many electrical applications such as spot-welding electrodes, seam welding of wheels and moulds for continuous and static casting. However, the thermal stability of UFG copper needs to be considered.

11.6 THERMAL STABILITY

Ultra-fine or nano-grained materials produced by any of the SPD techniques contain a huge amount of stored energy in the form of crystalline defects, such as dislocations and non-equilibrium grain boundaries. Most of the severely deformed materials exhibit low tensile ductility, and therefore, they must be annealed to obtain an optimum balance of strength and ductility. Furthermore, for long-term applications or further processing of the ultra-fine structured materials at temperatures above the processing temperature would inevitably cause unpredicted recrystallisation and it would lead to undesirable changes in the material properties. A transformation of the severely deformed state into a microstructure consisting of new recrystallised grains would therefore lead to a loss of mechanical property. However, the microstructure at least should retain its small scale during moderate annealing temperature in order to keep the high strength of the material. The temperature up to which the properties achieved by SPD is called as thermal stability.

Thermal stability can be assessed accurately to a large extent by its microstructural stability. The thermal stability can also be evaluated by measuring associated properties such as strength, hardness, electrical conductivity/resistivity, latent heat, etc. The stability of microstructure or other associated properties is usually determined by annealing at different temperatures for a pre-determined time or using differential scanning calorimetry (DSC). The thermal stability is useful in using the SPD materials for high-temperature applications.

UFG pure titanium prepared by CGP is subjected to annealing at different temperatures from 450°C to 600°C and for time durations ranging from 10 minutes to 6 hours. UFG titanium is thermally stable up to 450°C for 6 hours. The thermal stability of Al alloys containing 0.7% Li and 1.6% Li was studied using DSC. Thermal stability improved by the uniform dispersion of the second phase. A large number of fine secondary particles pins the grain boundaries and retards boundary diffusion. There are many influencing factors associated with the thermal stability of UFG/NS materials such as grain size/volume of grain boundaries, nature of grain boundary, dislocation density, latent heat (stored energy) associated with dislocations, elastic stress field, substitutional solutes, precipitates, segregation at grain boundaries, and many others.

Abnormal grain growth is found in CGP-ed low carbon steel annealed at 400°C and drop in strength and hardness values are observed for samples annealed at temperatures above 400°C. Regarding the presence of strain inhomogeneity in CGP-ed sheets, the driving force for restoration is not uniform and leads to dissimilar grain sizes. As a result, in the grain growth phenomenon that occurs at elevated temperatures, the kinetics of grain growth with dissimilar sizes leads to abnormal grain growth. Hence, 400°C is regarded as the optimum post-deformation annealing temperature for low carbon steel sheets to attain the minimum possible grain size and hence, best material properties. Similar results are also observed for copper, where CGP-ed copper sheets were annealed at 200°C, 300°C and 400°C. CGP-ed copper sheet

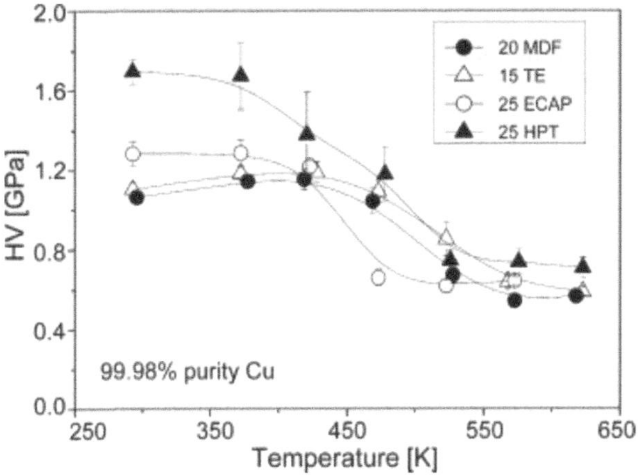

FIGURE 11.12 The hardness as a function of annealing temperature for Cu samples processed by 20 cycles of MDF, 15 passes of TE, 25 passes of ECAP and 25 revolutions of HPT.

annealed at 200°C exhibits new grains that resulted from recrystallisation. At 300°C, it exhibits some grain growth which is very insignificant. Annealing the CGP-ed copper sheets at a temperature of 400°C resulted in abnormal grain growth. CGP-ed copper sheets exhibited thermal stability up to 300°C. The hardness of the SPD-processed Cu samples measured after annealing for 1 hour at different temperatures between room temperature and 350°C is shown in Figure 11.12.

Despite the similar grain sizes (about 200 nm), the Cu samples deformed by various SPD methods showed different thermal stability at high temperatures. The thermal stability decreased in the order of MDF, TE, ECAP and HPT, which correlates to the increase of dislocation density and the decrease of the crystallite size. The UFG structure is maintained till 127°C (400 K) in all the methods. It can be understood that not only the initial grain size of the UFG materials but also the dislocation density and crystal size have a role in thermal stability.

In case of ECAP-ed CP-Ti processed in Route B_A, when annealed for 30 minutes at various temperatures shows that the microhardness decreases drastically when annealed at 450°C. While annealing between 300 and 400°C, a slight fall in hardness is noticed. Annealing samples between 300 and 400°C altered the irregular shaped grains to equiaxed and reduced the dislocation density significantly. Though the dislocation density decreases, it is still higher than the dislocation density in the annealed condition. This suggests that the cells and sub-grains with low-angle misorientation are formed during annealing. Steep fall in microhardness was observed when the samples were annealed above 400°C for half an hour due to recrystallisation. Thermal stability is also assessed based on the change in texture. In case of T40, a low alloyed titanium, when annealed at 500°C for different hours, it did not show any changes in texture and when annealed at 700°C, the texture changed to (10–13) and [11–20].

Thermal stability of ECAP-ed materials was studied by one in the authors' group on Al 2014 and structural steel and the results are summarised below: effect of processing routes of ECAP on thermal stability of 2014 Al alloy was studied in detail.

Al 2014 alloy is ECAP-ed in (i) annealed, (ii) solutionised and (iii) solutionised+aged in all routes. Up to five passes using a die with a channel angle of Φ =90° and a corner angle of ψ=20°. Isothermal annealing was conducted at various temperatures between 373 and 673 K in steps 373 K for 60 minutes in a muffle furnace to study the thermal stability. In case of ECAP-ed annealed Al 2014 shows gradual decrease in hardness is observed beyond the annealing temperature of 373 K, and a steep drop is observed at 473 K. With increasing temperature beyond 573 K, the softening of the material is accelerated, and there is a huge loss in hardness due to the dominance of recrystallised grains in the refined microstructure. Softening starts to occur at the same temperature of 373 K in all the processing routes. In the solutionised and aged condition, the softening behaviour is similar to that of the annealed condition. There is a continuous decrease in hardness after isothermal annealing beyond the temperature of 373 K. In solutionised and aged conditions, the hardness drops after 373 K is due to recovery, and beyond 473 K, the drop in hardness is more accelerated because of the grain coarsening in addition to precipitate coarsening in the microstructure. In the solutionised condition, the thermal stability is different from previous cases. The hardness drop is very less up to 473 K since the loss in hardness due to recovery is balanced by the formation of precipitates, and there is a very slight decrease in hardness beyond 473 K. Beyond 473 K, the softening occurs because of the coarsening of precipitates in the matrix. The processing routes in ECAP did not have any significant effect on the thermal stability. of the microstructure.

Similarly, the thermal stability of equal channel angular pressed structural steel IS2062 was studied. The ECAP-ed sample was annealed at varying temperature viz 500, 525, 550, and 600°C. For samples annealed at temperatures of 500 and 525°C, there was no change in the microstructure, whereas the strength and hardness slightly decreased, and the ductility improved due to the recovery process.

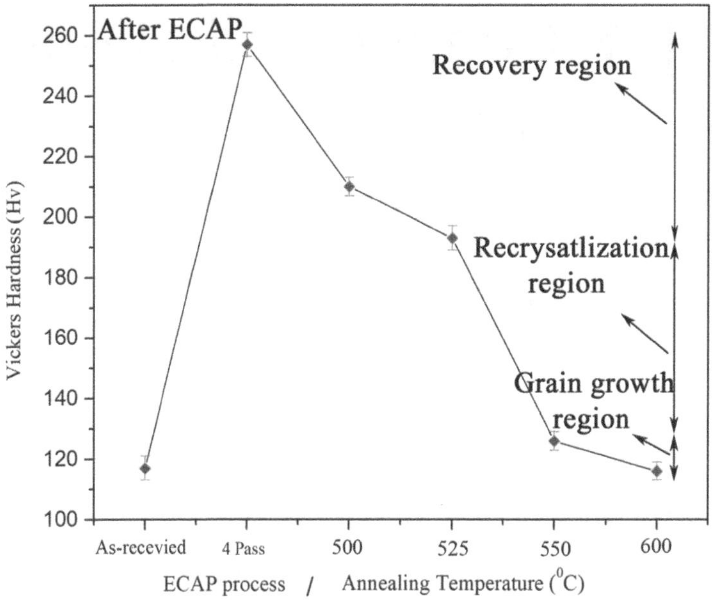

FIGURE 11.13 Vickers's microhardness evolution of annealed samples with respect to the annealing temperature.

The recrystallisation process occurred when the sample was annealed at 550°C. When the annealing temperature is 600°C, the grain growth process occurred in the ECAP-ed sample and the strength, hardness and ductility were almost equivalent to the base metal.

The microhardness of the sample at varying condition such as as-received, after ECAP and annealing, processed samples are shown in Figure 11.13, indicating the recovery, recrystallisation and grain growth stages.

12 Commercialisations and Challenges of Nanostructured Metals and Alloys

12.1 LIMITATIONS AND DEVELOPMENTS OF SPD PROCESSES TOWARDS COMMERCIALISATION

12.1.1 Equal Channel Angular Pressing

In the case of equal channel angular pressing, the major limitations are friction, complex geometries, manipulation between passes and the length of the product. Friction can be reduced using equal channel angular pressing (ECAP) dies with movable wall and rotatable rolls so that existing presses or equipment used for coarse-grained materials can be utilised for ultra-fine-grained (UFG) products. The complex geometries can be achieved by a profiled output channel. Multi-turn ECAP and multi-punch ECAP with different routes are used to manage the problem of multiple passes to reduce the production time. Due to these requirements, the design of the dies is critical. However, a well-developed tool/die design and making in the metal-forming sector can facilitate industrial implementation of ECAP. The limitation in length of the billet can be overcome by equal channel angular rolling (ECAR) or continuous ECAP, developed based on ECAP allowing repeated shear deformation on strip metals. ECAP is successfully commercialised for many materials, which are discussed in subsequent sections.

12.1.2 High-Pressure Torsion

The major constraint in commercialisation of high-pressure torsion (HPT) is that (i) the sample shape is in a form of disc which is not appropriate for many industrial applications compared to wire or sheet forms; (ii) the sample size is generally limited to a maximum of 35 mm in diameter in order to keep a high applied pressure; (iii) the strain is

DOI: 10.1201/9781003596110-12

generated in proportion to the distance from the disc centre and an inhomogeneous distribution of microstructure is developed across the diameter. However, these setbacks can be overcome by advancement in HPT processes. Continuous HPT enables the processing of any size sample without sacrificing pressure. However, CHPT is for producing metallic sheets and wires. Alternatively, incremental high-pressure torsion (IHPT) can handle cylindrical specimens. A combination of HPT and extrusion results in high-pressure torsion extrusion (HPTE), which can also handle cylindrical samples. Further, the strong velocity gradient in the specimen cross-section during HPTE provides the possibility to process hybrid materials or composite parts with functional elements. However, the HPT process needs further efforts for commercialisation.

12.1.3 Twist Extrusion (TE)

In TE, there are three obstacles for commercialisation: (i) sample size – needs to maintain aspect ratio to avoid buckling of sample and punches, (ii) friction and (iii) anisotropic nature of the fine-grained structure. Traditional extrusion and ECAP also possess limitations in sample size and friction. These limitations are overcome by the application of two-sided gripping of feedstock as well as containers and dies with movable walls. The same design concepts cannot be applied to TE, as the die channel is bound by twist, i.e., non-flat walls. TE with two parallel die walls, named Planar Twist Extrusion (PTE), has, therefore, been proposed to overcome these shortcomings. However, the aspect ratio is limited due to practical limitations. Under hydrostatic conditions applied through fluid can avoid the friction and the limitation in sample dimensions. Besides hydrostatic pressure leads to enhancement of sample workability, increase in maximum applied plastic strain before fracture, and a significant reduction in total load. Hydrostatic extrusion (HTE) solves the aforementioned issues and makes it suitable for the production of industrial-scale UFG samples.

Anisotropic nature of the Twist extruded samples can be minimised by changing the die profiles, increasing the operating temperature and subsequent recrystallisation heat treatment or by applying deformations like drawing or direct extrusion after processing. However, anisotropy cannot be eliminated completely. The alternative is combining TE with ECAP called as twist channel angular pressing (TCAP) combining the positive aspects of both into a unique novel method. Though the TE is more suitable for commercial production, till date no industrial product is available commercially.

12.1.4 Cyclic Extrusion and Compression (CEC)

As compared to the other SPD methods, CEC is more suitable for industrialisation as it imposes a large deformation leading to intense grain refinement in a single pass. The other advantage is the application of triaxial compressive stress and high hydrostatic pressure due to back pressure, which prevents billet failure, enabling the processing of many hard-to-deform materials such as hexagonal-close-packed (hcp) metals and alloys at relatively low temperatures. Also, repeating this plastic deformation in order to implement high plastic strains does not require removing the billet at intervals of consecutive passes. Therefore, large plastic strain can be attained without removing it from the die during the successive passes, which is another benefit of this method compared to other SPD techniques. Regardless of the many advantages that the CEC method has, unfortunately, has some limitations due to non-uniform strain distribution that causes non-uniform microstructure and resultant non-uniform properties and geometry, size, and shape of the samples.

The recent modifications in CEC methods are encouraging in the production of long-size billets that can be used readily on an industrial scale. One such is the Hydrostatic cyclic extrusion compression (HCEC) process, which can produce UFG billets without any restriction on the dimensions of the cross-section area and length of the billets. In addition, since a pressurised fluid encompasses the billet during the process, it eliminates the frictional forces, thus significantly decreasing the required processing force. In a similar way, Hydrostatic cyclic expansion extrusion (HCEE) can also produce quite long length billets. In addition, this process can handle asymmetrical components. The non-uniform strain distribution can be minimised by combining the ECAP process, called the cyclic extrusion channel angular pressing (CECAP) process. Besides, tube cyclic extrusion compression (TCEC), Hydrostatic tube cyclic expansion extrusion (HTCEE) can process tubes and thus overcome the geometry limitation of the samples. Apart from processing samples for UFG structure, it is useful in consolidating powders and chips for solid-state recycling of alloys and composites, which can be commercially used.

12.1.5 Multi-Axial Forging

Multi-axial forging (MAF) is one of the SPD processes that can be easily commercialised as it balances cost, speed, and quality among the SPD processes. The main advantage of MAF is its simple tool design and ability to process bulk materials. The main limitation is the inhomogeneity due to the localised deformation along the direction of shear resulting in inhomogeneous microstructure across the cross section at initial passes. However, increasing the number of MAF passes can partially help to solve the problem, but arbitrary increase in the number of passes may lead to redundant increase in manufacturing cost and time. The other major limitation is effectively controlling the temperature. Uneven distribution can cause problems such as hot tears or weak spots in the material. This is especially true when working with large parts where the heat may not spread evenly throughout the component. In multi-axial diagonal forging (MADF), requires less frequent mould replacement and induces relatively high levels of hydrostatic pressure inside the material that suppresses the generation of forging defects even when the number of cycles increases. The heterogeneity is less in MADF. Though MAF is more suitable for commercialisation, no commercialised product has been developed so far.

12.1.6 Accumulative Roll Bonding

Accumulative roll bonding (ARB) is one of the SPD processes for sheet metals. As the load is applied through rolls, the process has no limitation in length of the product. It is applicable not only for SPD of materials but also for the production of dissimilar metals and for the production of nano-composite materials. The major limitation is the temperature of deformation since the bonding is involved. Low temperatures can result in insufficient bonding and bond strength, while high temperatures can cause grain coarsening and dynamic recrystallisation. With a little sacrifice in dimensions, cryo rolling followed by ARB solves this issue. The other problem is mechanical properties, especially the hardness of ARB sheets, which can fluctuate greatly due to the interface and uneven deformation between sheets. Precise control over deformation and temperature is required, which makes ARB relatively expensive. ARB is not mature enough so far for commercialisation; however, the development of precise sensors to control the process parameters brightens the possibility.

12.1.7 Groove Pressing (GP)

GP is another SPD technique for sheet metals. The major advantage as compared to ARB is that there is no repetitive bonding between sheets, which enables performing GP at lower temperatures as compared to ARB. Comparing the GP and ARB processes based on experimental results, the final grain size of alloy obtained after three passes of GP is smaller than the grain size for the same alloy subjected to eight passes of ARB. The major limitations are (i) due to the sharp profile of the dies, micro-cracks appear on the surface of the specimen, and they lead to a decrease in strength and other mechanical properties, (ii) inhomogeneity in strain distribution and (iii) dimensions of the sample.

Rubber pad-constrained groove pressing (RP-CGP) overcomes the first problem by replacing the upper die and punch with a flexible rubber pad, which not only eliminates one set of die and also induces the hydrostatic condition, eliminating cracking of the material. Hence, the sheet material can be subjected to a greater number of passes. In addition, constrained groove pressing-covered sheet casing (CGP-CSC) where the sample is sandwiched between two other sheets of hard material such that the combined thickness of all the sheets is equal to the groove width. The sheets of hard materials are in contact with the sharp grooves, preventing the formation of micro-cracks on the sheet's surface. Hence, this method ensures good surface quality for the specimen when compared to the conventional CGP process. Groove pressing-cross route (CGP-CR), eliminates the second problem by rotating by an angle of 90° around axis perpendicular to its plane such that the RD is now perpendicular to the die length between each pass, which makes a more uniform strain distribution in the sheets. Continuous GP machine is capable of producing bulk nanostructured materials where sample dimension (length) is not limited by the length of the die.

These advancements brighten the possibility of sheet metals with ultrafine/nanostructured sheets for commercial production.

12.1.8 Friction Stir Processing

Friction stir processing (FSP) is the best method to alter the surface properties commercially as it is not limited by the shape and size of workpiece and the processing environment. The major limitations include:

The need for precise clamping and tooling, precise control over potential for defects like "worm holes" if parameters are not optimised, limitations in complex geometries, relatively slow processing speeds, high tool wear, and the requirement for specialised operator skills. The specialised machinery required for FSP is expensive, impacting the overall cost of production. These limitations make it less suitable for large-scale production.

However, FSP is the most valuable process for the production of surface composites. The recent advancement in the incorporation of reinforcement such as direct friction stir processing, water cooled FSP, cryogenic friction stir processing and friction stir vibration processing are showing the way for commercialisation for critical applications.

12.2 COMMERCIALISED NANOSTRUCTURED METALS AND ALLOYS

At present, more demanding commercialised nanostructured materials produced by SPD techniques are in the area of bioimplants. The most preferred bio implant for many applications is Ti-6Al-4V ELI grade. Though titanium is biocompatible, the alloying elements Al and V are not. If pure titanium is used instead, toxicity can be avoided but strength and endurance limit of pure Ti is much less than that of Ti-6Al-4V. However, some implants that require less fatigue life are made with SPD CP-Ti. Among the different approaches tried during the past two decades for improved biocompatibility along with better mechanical properties, nanostructuring of metals and alloys by severe plastic deformation (SPD) has shown promise since the processing methods of SPD form nanostructures not only within bulk metals and alloys but also on their surfaces, which makes nearly 70% of medical devices that are in use. The commercially produced metals by SPD are used for trauma, orthopaedic and dental applications.

One such application is bone trauma fixation plates. A mini-plate for bone fixation made of CP-Ti, as specified by the ASTM F67 consensus standard, was used as the starting point for redesigning the product dimensions for making a nanostructured CP-Ti version of the mini-plate. Due to the higher strength of the nanostructured CP-Ti, the plate thickness is reduced from 0.9 mm to 0.7 mm. The other most commonly used implant is for dental applications. Dental implants need to withstand high loads associated with chewing or clenching of the jawbone during stressful sleep. The conventional implants are of size 2.9 mm or greater. The nanostructured Ti Grade 4 of lesser diameter (2.4 mm) produced by the ECAP-Conform (Continuous ECAP), shows very good clinical performance because of the added strength due to nanostructuring. Figure 12.1 shows some of the implants made out of SPDed CP-Ti grade 4.

Apart from bio implants, nanostructured filler wires incorporated with nano particles is used in fusion welding for metals/alloys susceptible to hot cracking. One best example is the welding of high-strength aluminium alloy AA7075, which has been considered difficult to join by fusion welding since its invention in the 1940s due to its hot crack

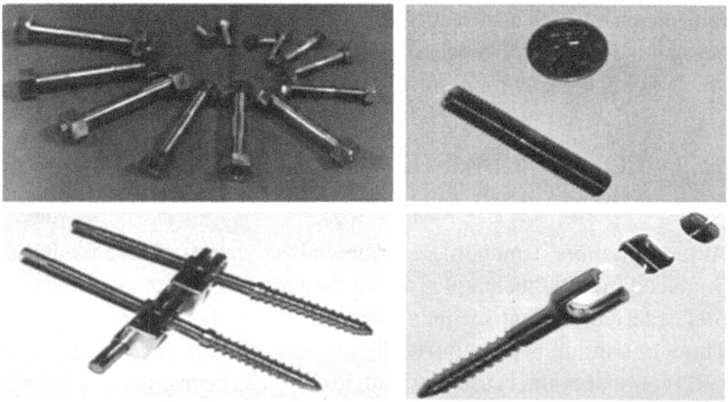

FIGURE 12.1 Bio implants made out of SPDed CP-Ti for commercial applications.

susceptibility, and is safely arc welded without hot cracks using nanostructured 7075 aluminium filler wire containing TiC nanoparticles produced by SPD processes in combination with wire drawing.

Recent publications are in focus of producing metallic conductor wires with high mechanical strength and good electrical conductivity for electronics to high-voltage power transmission cables, especially in railways. The alloying addition to the metallic conductor materials, such as copper and aluminium, in a view to improve the mechanical strength decreases electron mobility considerably. The alternate way is in the production of nanostructured copper and aluminium to enhance conductivity, strength, and other properties. The nanostructured materials produced by the SPD process have found commercial application in electric motors, high-speed railway contact wires, heat exchangers, and electric resistance welding electrodes. Therefore, high-strength and high electrical conductivity are required. The results are encouraging. Increase in 5% IACS electrical conductivity has been reported for ECAP of AA6xxx alloys. At the same time, the strength is also increased. The substitution of nanostructured aluminium conductors for copper in automobiles is another convincing example because of the 1.65-fold weight savings per delivered amp, the 1.5-fold lower cost per amp, and the enhanced fuel economy associated with the weight reduction. Furthermore, reducing the amount of copper wiring in automobiles increases the extent to which automobile steel can be recycled without contamination from copper.

Companies in Europe, North America, and Asia are exploring the advantages of nanostructured conductor alloys. The other one is Cu–WC nanocomposites, which exhibit enhanced hardness/strength, Young's modulus, and thermal stability while maintaining high electrical/thermal conductivity. For example, such a high level of properties of more than 600 MPa in tensile strength and 85% IACS in electrical conductivity is achieved.

The important point is commercialising these alloys for applications demands high volume, continuous length, and it should be available in the form of coils. Recently, coil-to-coil C-ECAP has been demonstrated for low-melting temperature alloys, including aluminium and magnesium. Coiling has been applied to produce AA6101 and AA6201 conductor alloys up to 12 mm in diameter. High-speed operation of C-ECAP has been piloted, and these results have been used to estimate annual production volumes within a large-scale manufacturing environment, including consideration of alloy yield efficiency, tooling changes, coil handling, and machine maintenance. The achievement of both higher electrical conductivity and strength is attractive for applications such as overhead power transmission cables or replacing copper wiring in automobiles with aluminium.

Low-carbon advanced nanostructured steels can serve various structural engineering applications, from bridges and automobiles, to stress-critical applications such as reactor pressure vessels in nuclear power stations, since steels are among the lowest cost of all alloys. There are three major types of nanostructured steels that have been successfully commercialised. Out of these three types, nanostructured advanced high-strength steel (AHSS) is produced by a combination of heat treatment and SPD to achieve the refinement of grains and secondary phase. However, due to the necessity of SPD, the nanostructured AHSS can only be produced in steel sheets or rods. The other two types, namely nanocrystalline bainitic steel and oxide dispersion-strengthened (ODS) steel, which are produced by solid-state phase transformation and by ball milling.

The superior properties of nanostructured alloys, mainly for medical applications, offset the higher prices of such materials. Other potential price-insensitive applications include luxury jewellery, where SPD-processed materials offer higher strength, hardness, and better wear resistance.

12.3 CHALLENGES AND OBSTRUCTIONS IN COMMERCIAL APPLICATIONS OF NANOSTRUCTURED MATERIALS

Before a process can be commercialised. It should be established that there are significant market demands. Once demand is created, a range of factors must be considered that can be an obstruction for commercialisation. This section discusses these factors briefly.

The primary factor is the integration of the nanostructuring process with other steps in conventional manufacturing to get the product economically. The integration has yet to be demonstrated for nanostructured metallic conductors, which is to some extent established in bio implants. The second one is the long-term stability. Thermodynamic driving forces to reduce free energy in all material systems favour the gradual elimination of the same microstructural features that underpin the superior properties of nanostructured metals. Reduction of grain-boundary area, decrease of dislocation densities, and elimination of nonequilibrium structures can result in issues with long-term stability. Only after many years of documented performance can nanostructured metals be considered for some applications, such as long-term medical implants, electrical conductors, etc. Thus, the adoption of nanostructured metals will be deliberately delayed for applications in which long-term stability must be known in advance.

The other setback is the characterisation of nanostructured materials. Advanced characterisation tools such as atomic force microscopy, three-dimensional atom probe tomography, or electron backscatter diffraction by transmission Kikuchi diffraction partly enable the development of nanostructured metals and alloys. The use of such techniques is problematic and uneconomical for commercialisation. Manufacturers of metal products invariably need to inspect the metals and alloys they receive. This can be problematic when inspection of nanostructured metals requires advanced characterisation techniques. Even a basic inspection of standard material parameters, such as grain size, typically performed via optical microscopy for conventional metal are ineffective for nanostructured metals since the grains can be too small to be resolved with white light. Furthermore, the equipment needed to characterise nanostructured metals are expensive and are rarely available in manufacturing environments. Also, personnel with the level of education (graduate level) needed to operate and interpret data from advanced characterisation tools, such as electron microscopes, are often unavailable in manufacturing environments. Consequently, it can be difficult to adopt nanostructured metals in commercial applications without investing in high-end characterisation equipment and hiring experienced scientists with graduate degrees. The personnel and equipment needed to conduct the elaborate techniques to document microstructures upon receipt of nanostructured metals are beyond the reach of many industries. Not only the microstructure, but even the mechanical testing of nanostructured metals and alloys is problematic. In many industries, hardness tests are used for routine check-ups, and the hardness can be easily correlated with conventional materials. However, hardness and tensile strength do not correlate as consistently for nanostructured alloys. Thus, this established, quick method to approximate tensile strength is not as useful for nanostructured metals.

The other obstacle is the availability of standards. In case of conventional materials and methods, consensus standards such as ISO, DIN, ASTM, ANSI, and others are widely available. Yet, for some nanostructured alloys, standards do not yet exist. Consequently, products for highly regulated industries such as aerospace and medical devices are slow to adopt nanostructured metals, unless they are adequately covered by existing standards. While standard developing organisations worldwide have formed nanotechnology-related committees to review and approve standards for nanoscale engineered materials and products, their slow emergence lags the industries that are ready to adopt the benefits of nanostructured metals and alloys.

The most significant barrier is the disbelief about the improvements in properties in nanostructured material, which can be 30%–200% as compared to existing products which creates a barrier to their adoption. The other factor is the regulations specified governments with respect to safety, environmental protection, and energy efficiency must be met by the materials produced by SPD processes which delays the commercialisation processes. However, some regulations, for example, the Corporate Average Fuel Economy (CAFE) standards in the United States and similar regulations in European Union, create a motivation for rapid adoption of nanostructured metals. In case of bio implants, any substitution of a new material requires extensive validation to prove that the substitution results in improved performance to achieve regulatory approval. Thus, new nanostructured alloys to conventional alloys undergo comparable scrutiny. However, nanostructured alloys may have an advantage in that alloys with previously approved chemical composition can be nanostructured to enhance performance, without the additional risks posed by introducing new chemical compositions. This simplifies the regulatory approval processes. Similarly, simple single-phase alloys such as Ti-15 wt.% Zr can be nanostructured to achieve high strength. However, there is a change in composition and hence they are less attractive for medical applications.

Another area in which greater knowledge is needed by consumers and manufacturers is the cost to purchase and use nanostructured metals and alloys. Because additional steps are required to impart fine-scale structures in metals, it is appropriate to expect increased costs to manufacture nanostructured materials. However, other factors may offset higher production costs. For example, studies have shown that nanostructured metals have greater machinability and formability than their conventional counterparts. Thus, the total cost to manufacture products for which extensive machining is required may be less by using nanostructured alloys.

There is thus a need for educating consumers on these characteristics, especially since the advantages and limitations of nanostructured metals differ between alloy families. Generalised claims of benefits of nanostructuring have the potential to mislead prospective adopters of nanostructured metals. Specific and easily understood evidence of the advantages and disadvantages must be available to inform nonexperts about nanostructured metals and alloys.

Index

Q

R

S

T

Z

For Product Safety Concerns and Information please contact our EU representative GPSR@taylorandfrancis.com Taylor & Francis Verlag GmbH, Kaufingerstraße 24, 80331 München, Germany